"ONE STOP BEYOND DORKING..."
A Holmwood Station Scrapbook

At approximately 10.40 am on 13th June 1965, 'Battle of Britain' class pacific No. 34050 'Royal Observer Corps' thunders through Holmwood station whilst working 'The Wealdsman Rail Tour', organised by the Locomotive Club of Great Britain. Although the goods yard has recently been lifted, the station building is still intact, complete with the original covered access stairways to both platforms. To the left is the layover siding used by electric multiple units that once terminated at this station and the cat-walk that allowed their crews to change ends. Happily, trains hauled by steam locomotives can still be seen and photographed at Holmwood.

"One Stop Beyond Dorking..."

A HOLMWOOD STATION SCRAPBOOK
to celebrate 150 years in our community

1867 – 2017

Written and compiled by

Julian Womersley

Published by YouCaxton Publications

First published in 2017

Copyright © Julian Womersley 2017

The moral right of J S Womersley to be identified as the author has been asserted by him in accordance with the Copyright, Designs & Patents Act 1988

All rights reserved.

No part of this publication may be reproduced, stored in a retrieval system or transmitted in any form or by any means, including electronic, electrostatic, magnetic tape, mechanical, photocopying, recording or otherwise, without prior permission in writing from the copyright holder and publisher.

ISBN 978-191117-56-81

For photographic and other credits, please see page 138.

Contents

BEFORE THE RAILWAY	1
SOME FALSE STARTS	3
THE HORSHAM, DORKING & LEATHERHEAD RAILWAY	9
TRIALS AND TRIBULATIONS DURING CONSTRUCTION	11
A DIGRESSION ON COMPULSORY PURCHASE	17
THE EARLY YEARS OF HOLMWOOD STATION	24
'THE BATTLE OF DORKING' AND ITS AFTERMATH	30
BETCHWORTH TUNNEL: ITS COLLAPSE & OTHER MISADVENTURES	33
EPISODES FROM LATE 19th CENTURY COUNTRY LIFE	38
HUNTING & THE RAILWAY	44
HOLMWOOD JUNCTION? THE HOLMWOOD & CRANLEIGH RAILWAY	48
THE DEATH OF QUEEN VICTORIA	50
EPISODES FROM EARLY 20th CENTURY COUNTRY LIFE	56
MORE ROYAL TRAINS AND OTHER SPECIALS	64
SUFFRAGETTES AND A SOCIAL CONSCIENCE	68
THE GREAT WAR AMBULANCE TRAINS	74
HOLMWOOD SIGNAL BOX	78
BETWEEN THE WARS	85
THE GOODS YARD	91
INTO ANOTHER WORLD WAR	98
STATE INTERVENTION	102
STEAM SPECIALS	116
EPILOGUE	122
	124
	126
	128
	134
	135
	136
…DGEMENTS	137
…PHIC & OTHER CREDITS	138
…GPHY	139
…RCES	140
…E AUTHOR	141

PREFACE

Today, Holmwood station is but an unmanned halt, with a limited service and no trains on Sundays. After decades of corporate neglect, there is no doubting that its looks are faded now and most of the original buildings have been demolished. Yet, as will become clear, this is a railway station with a remarkable past. It has witnessed astonishing events and extraordinary people have trodden its platforms.

It is self-evident that for a railway station to exist there must first be a railway. The struggle to get a line built across this part of Surrey took the best part of forty years, but that tale is worth telling - particularly as one of the doomed schemes intended to adopt that fleeting marvel of the Victorian age, the atmospheric railway. Indeed, whilst these various failed enterprises may have been catalogued before, I am not sure that any detailed research into them has previously been published. The hallmarks of that particular saga, the optimism and disappointment of the unsuccessful ventures, together with the final triumph of the Horsham, Dorking and Leatherhead Railway in 1867, are all described using various contemporary newspaper reports.

Once built, the railway was absorbed into the wider system of the London, Brighton & South Coast Railway and very quickly the line started to make its mark on the travel, commercial and social habits of the local population. The sheer variety of these new opportunities soon becomes apparent from reading the coverage in the local newspapers of the comings and goings of rich and poor alike at Holmwood station.

Pauper children passed through on their way from inner London workhouses on their way to foster homes on Holmwood Common, whilst in the other direction well-heeled gentlemen, including a Crimean War hero and the maker of the modern Lloyd's of London, made their way, first class, up to 'Town'. Heroes returned from the Boer War and, during the Great War, wounded soldiers straight from the Western Front were taken on stretchers from ambulance trains in the dead of night. In the goods yard, coal and other essentials came in and livestock, agricultural produce and other goods, including distress rockets and flares, went out. Special trains came and went - bringing London society people to glittering parties in the country, taking Sunday school outings to the seaside, the Primrose League to the Crystal Palace and transporting the militia to camps and military exercises on Holmwood Common. Railway employees, including the first two station masters, were dismissed from their posts for financial irregularities, others for insobriety. There have been accidents at the station and people have died. The station was also frequently used by suffragettes and, surprisingly, the Surrey Union Hunt, who unloaded hounds and horses directly onto the platform. It has also appeared in at least two works of fiction.

But perhaps the most improbable of all the many people to alight onto the up platform over the years was His Imperial Highness, Emperor Wilhelm II, the German Kaiser. He was accompanied by His Royal Highness, The Prince of Wales on his way to be proclaimed His Majesty, King Edward VII and His Royal Highness, The Duke of York. Other Royal trains have also passed through Holmwood, including that used for the funeral of Queen Victoria.

Delving into history can be a chancy business, often undertaken in the expectation that sufficient material will become available to provide a seamless, well-balanced and comprehensive narrative. However, in this particular instance, it soon became clear that whilst there was a glut of information to be gleaned from some periods, in others there was an almost complete dearth of detail. Moreover, I greatly prefer steam locomotives to electric multiple-units! So this work makes no claim to be the definitive history of Holmwood railway station over the years. Instead, perhaps, it should be viewed more as a serendipitous collection of snippets; photographs; extracts from archive documents; press cuttings and other material grouped together and allowed to tell their own story - just like a scrapbook, in fact.

Whilst this is a book on a railway subject, it is written for those interested in social history and trainspotting alike. Hence whenever technical matters appear, I have tried to explain them in their appropriate context. The only obvious exception to this rule is the use of the words 'up' and 'down' as adjectives - the former should be taken to mean 'towards London', whilst the latter implies 'away from London'.

Finally, the title: as so very few people have ever heard of Holmwood station, the answer to the question, "Where do you commute from?", is inevitably followed by, "Oh, where's that?". Hence my stock reply, "One stop beyond Dorking ...".

Julian Womersley
Beare Green
[formerly known as Holmwood Station]
Near Dorking, Surrey
March 2017

A Note on Spellings, Money & Measurement:

Spellings: The surveyors who produced the various plans shown in the text recorded the names of settlements and farms as they heard them whilst tramping across the countryside. These phonetic transcriptions have led to a variety of spellings, notably for Bregsells Farm; Breakspear Farm; Ockley and Dorking. Rather than viewing these as quaint misapprehensions, readers are encouraged to recreate the vowel sounds of the rural Surrey accent of the early 19th century by reading the various names aloud, but in the knowledge of modern spelling and pronunciation. Such words are shown in italics.

Money: Prior to decimalisation on 15th February 1971, English currency was divided into pounds (£ or ℓ), shillings (s. or /-) and pennies (d.). There were 20 shillings in £1. A shilling comprised 12 pennies. A penny was subdivided into 2 halfpennies or 4 farthings. Other coins were the three-penny bit, 'thruppence' [3d]; the six-penny piece, 'a tanner' [6d]; the 2 shilling piece or florin [2/- or 24d]; the half crown [2/6] and the crown [5/-]. Gold coins included the sovereign [£1] and the half sovereign [10/-]. Although last struck in 1799, the guinea [£1/1/0] and the half guinea [10/6] were considered a more 'gentlemanly' form of fiscal exchange. Hence tradesmen rendered their accounts in pounds, whilst fine art or race horses were priced in guineas.

Measurement: Only Imperial units of measurement appear in this book. Linear measurement is based in the statute mile, containing 1760 yards. A yard has 3 feet, each of which is comprised of 12 inches. All the surveys illustrated were made using a Gunter's chain, which is 22 yards [or 66 feet] long. A furlong is 10 of these chain lengths. Thus there are 8 furlongs, or 80 chains, in 1 mile. A chain is comprised of 100 links and 25 links comprise a rod, perch or pole. Thus there are 4 rods, perches or poles in a chain, each having a length of 5½ yards [ie 22 yards ÷ 4]. Square measurement is based on the acre, containing 4840 square yards. In an acre there are 4 roods, each comprising 40 square rods, perches or poles. Thus 1 square rod, perch or pole = 30¼ square yards and 1 rood = 1210 square yards. Happily, later in the 19th century, the polar planimeter allowed the use of decimal fractions in area calculations!

Chapter 1

BEFORE THE RAILWAY

Until the middle part of the 18th century, travelling between Horsham and Dorking was virtually impossible during wintertime because the heavy and impermeable nature of the Wealden clay made roads and tracks into a quagmire. Even in summertime roads were rough and rutted, making it difficult for local producers of goods and agricultural merchandise to gain access to the lucrative markets of the metropolis. This sorry state of affairs was ameliorated following the creation of the turnpike between Horsham and Epsom in 1755.

On the right of this page, an extract from the 'Direct Roads' map by Carington Bowles, dated 3rd January 1785, shows the route of this turnpike between Leatherhead [to the north] and Capel [to the south]. The sparsely populated nature of the countryside between *Darking [sic]* and Capel is most noticeable. A second turnpike arrived in *Bear Green [sic]* some years later from the *Okeley [sic]* direction. The ensuing traffic soon being sufficient to encourage the establishment of a smithy at their junction and two inns to the north of it: the 'Duke's Head' and 'The White Hart', respectively.

Further north still, beyond the toll-house near the 27 milestone at Holmwood Corner, the beauty and quietude of the Surrey Hills were starting to be recognised by incomers. On *HomeWood Common [sic]*, erstwhile commoners' cottages were being 'gentrified' and a small residential enclave started to grow around the church of St Mary Magdalene which had been consecrated in June 1838.

This residential expansion encouraged the development of several brick and tile works in the vicinity and various trades-people to set up businesses serving the growing population. Similarly, farming was becoming more prosperous and reliant on transport to bring in supplies of seed and fertilizer or to take agricultural products to market.

Located between the two thriving market towns of Dorking and Horsham, the neighbourhood around Holmwood was becoming ripe for a railway speculation to further improve its transport links not only to them, but also to London and the rest of the country.

However, despite the mania for railway investment sweeping across the rest of Britain during the mid-nineteenth century, in the end it took quite some time for a railway to finally arrive in Holmwood. Yet, and let there be no misapprehension on this point, it certainly wasn't for the want of trying. A succession of promoters appeared on the scene, all of them keen to see their ambitious ventures connect the growing towns of Leatherhead, Dorking and Horsham.

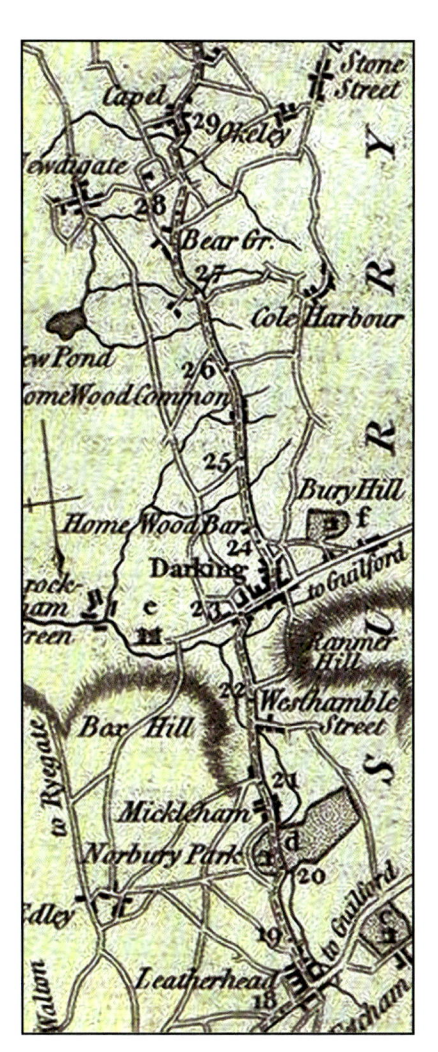

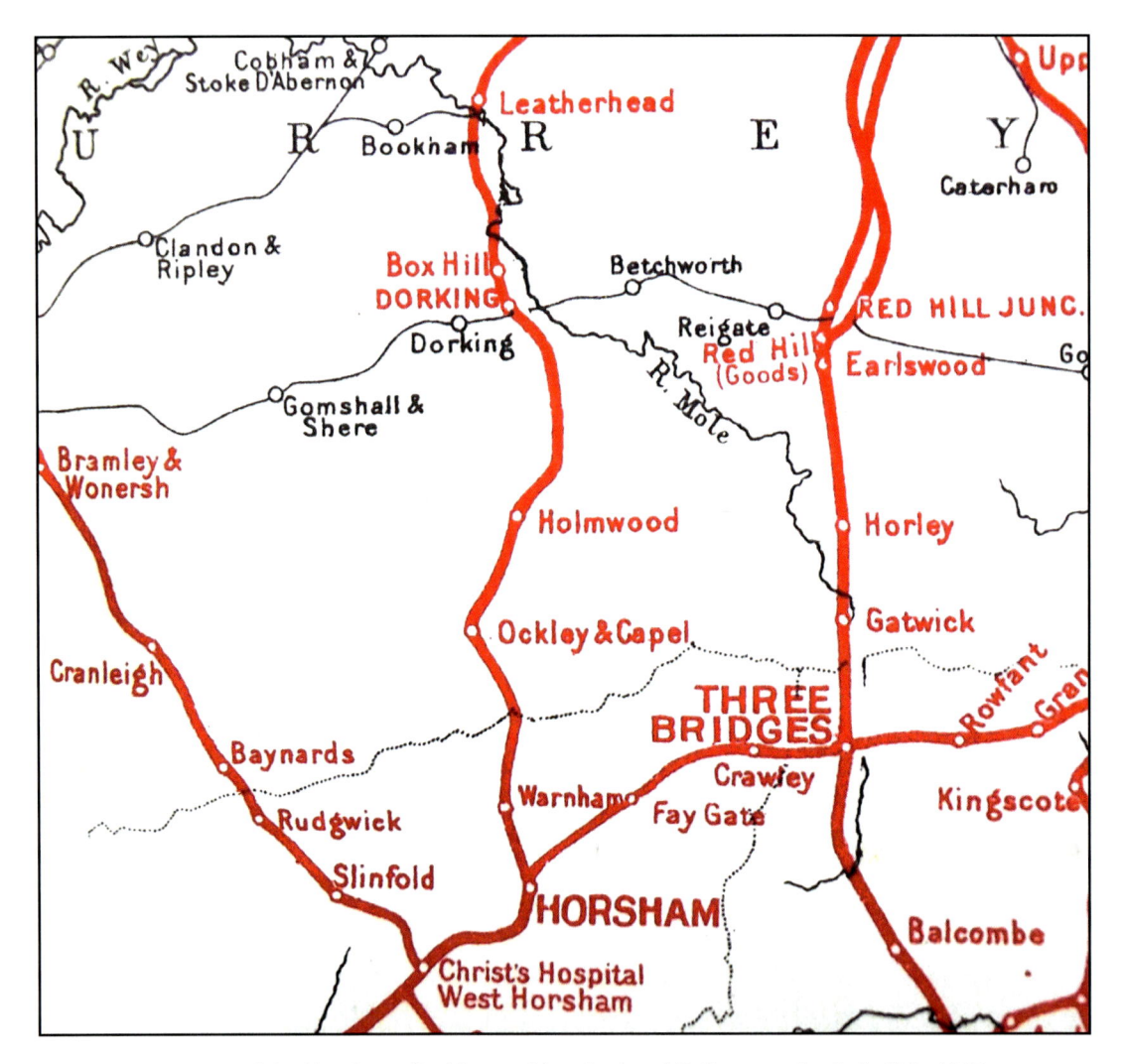

The route of the Horsham, Dorking and Leatherhead Railway, as finally built in 1867.
The length of the line from Horsham to Dorking is 13 miles 24 chains and from Dorking to Leatherhead it is 3 miles 78 chains.
Between Dorking and Holmwood the eastward loop around the relatively high ground of Holmwood Common is very prominent.
The SE&CR railway [1849] between Redhill Junction and Guildford, via Reigate and Gomshall & Shere, runs East-West across the map.
The LB&SCR direct route to Brighton [1841] runs through Redhill Junction, Gatwick [for the racecourse] and Three Bridges.
The LB&SCR Horsham to Guildford line [1865] runs via Christ's Hospital and Cranleigh.

Chapter 2
SOME FALSE STARTS

Probably the earliest proposal for a line running through the Mole Gap between Leatherhead and Dorking and on, broadly vîa Horsham and Shoreham, was in 1829 when Charles Blacker Vignoles surveyed this option at the invitation of John Rennie, who had been commissioned to examine several possible railway routes to Brighton. This ambition had to be abandoned due to a lack of support in Parliament.

In 1835, this route was again to be used by two of the several competing designs put forward in an attempt to capture once more the potentially lucrative passenger traffic between London and Brighton. Indeed, it was this traffic that was the driving force, rather than any desire to serve intermediate towns on the way, the two routes through Dorking being explored solely on account of engineering convenience. One of these rivals was the Grand Southern Railway, proposed by Nicholas Wilcox Cundy, to run from Nine Elms, vîa Mitcham. The other was Robert Stephenson's route, branching off from the London and Southampton Railway at Wimbledon, vîa Epsom. As both of these proposed lines were some 13 miles longer than the direct route between London and Brighton, vîa Merstham and Haywards Heath, it is not surprising that the shorter option was eventually adopted and opened in 1841.

It was a branch from Three Bridges, a station on this London and Brighton line, which first brought the railway to Horsham on 14th February 1848.

In Dorking, the first railway in the town arrived at what is now Dorking West station on 4th July 1849, following the opening of the section of the Reading, Guildford & Reigate Railway from the Reigate Road junction [later to be called Redhill] with the South Eastern Railway's main line from London Bridge to Dover. The westward extension of this line, to Guildford and beyond, opened later that same summer.

An Epsom and Leatherhead railway was authorised in 1856, but it was not until 1859 that the railway finally reached the latter town.

Thus, by 1860, whilst the towns of Horsham, Dorking and Leatherhead all had links to the growing national network, they still did not have a direct railway connection running between them in a north-south direction - albeit, as the increasing number of failed schemes demonstrates, this was not due to any lack of enterprise or ingenuity.

Of these, undoubtedly the most imaginative was the proposal for a Dorking Brighton and Arundel Atmospheric Railway. A prospectus for the venture was issued in 1845 and, in the autumn of that year, the 'Brighton Gazette' of Thursday, 16th October carried a lengthy notice setting out the intentions of the company, naming its supporters, the route of the line and how to apply for shares. A slightly abridged version of this notice is set out as Appendix: 1.

Perhaps needless to say, these intrepid investors were soon to lose all their money on this venture.

With the benefit of hindsight, the choice of atmospheric traction may seem an odd one. However, in the 1840s, using a modified natural air pressure in an iron tube laid between the rails to move a piston attached to a train was an exciting prospect. In theory, it could offer a cheaper, faster, safer and cleaner alternative to locomotive hauled trains, especially on hilly routes where extra assistance had

THE DORKING, BRIGHTON AND ARUNDEL ATMOSPHERIC RAILWAY.

By Horsham and Shoreham, without a Tunnel
(Provisionally registered according to Act of Parliament)
Capital, One Million, in 50,000 shares of £20 each.
Deposit. £2 2s. per Share

ARRANGEMENTS are in progress for the formation of a Company to make an Atmospheric Railway from Dorking (or Epsom) to Brighton, through Horsham and Shoreham, and from Horsham to Arundel, thereby affording (either by means of a junction with the Direct London and Portsmouth Atmospheric Railway at Dorking, or the Croydon Atmospheric Railway at Epsom) to the intermediate very populous and wealthy district the important advantage of a direct communication with the metropolis, and also with the coast from which it is at present excluded, as well as forming the best connecting link between that district and the Ports of Shoreham, Arundel, Portsmouth and Southampton, and the principal Towns of Surrey, Sussex, Kent and Hants.

A detailed Prospectus with the names of the Provisional Committee will be shortly published; and in the meantime communications may be addressed to Messrs. Campbell and Witty, 21, Essex Street, Strand; Messrs. Attree, Clarke and McWhinnie, and Messrs. Upperton, Verrall and Veysey, Brighton; Messrs. Coppard and Rawlinson, Horsham: Messrs. Everest and Wardroper, Epsom; and Richard Holmes,Esq., Arundel.

CAMPBELL and WITTY
Solicitors for the Bill

The notice that appeared on page 2 of The Brighton Gazette on Thursday, September 11th 1845, announcing the launch of the Dorking, Brighton and Arundel Atmospheric Railway scheme.

An atmospheric train in action - a piston carriage on the London to Croydon Railway. The rather alarmed-looking driver is screwing down the brake, whilst behind him [to the left] is a vacuum gauge. The cast iron tube carrying the partial vacuum that powers the train lies underneath the carriage, between the running rails.

to be provided because adhesion could still be a problem for the heavier trains on gradients over 1 in 75. The system used steam-powered pumping stations built at strategic locations along the railway to create a partial vacuum within the pipe. The reduced pressure produced a sucking effect in front of the piston which, in conjunction with normal atmospheric pressure behind, was sufficient to propel a train attached to it.

The first commercial atmospheric railway to offer a public service was that operated from 29th March 1844 to 12th April 1854 between Kingstown and Dalkey by the Dublin & Kingstown Railway. Nearer to home, the London & Croydon Railway ran an atmospheric line over the 5 miles between Dartmouth Arms [later Forest Hill] and Croydon [now West Croydon] from 19th January 1846. Although this railway was extended by a further 2½ miles from Forest Hill to New Cross in February 1847, all services ceased on 3rd May 1847.

So, when the Dorking, Brighton and Arundel Railway Atmospheric Railway was first mooted, it did seem to prospective backers of the scheme that there was a chance that this novel form of propulsion might just succeed, especially as one of its supporters was the famous engineer, Isambard Kingdom Brunel. However, there was some powerful opposition to the proposal, with the Duke of Norfolk making his views particularly plain in the press:

> **The Sussex Advertiser, Tuesday 7th October 1845**
> *Lewes* – The Duke of Norfolk has publicly avowed his intention of opposing the Dorking, Brighton and Arundel Atmospheric Railway "in every way" he "may think most likely to be successful".

The formation for this line, surveyed by Charles Vignoles, was to have run from Dorking, across Meadowbank and over the Guildford Road, demolishing the Parsonage House in the process. Staying to the West of Dorking before eventually swinging over the Dorking to Horsham turnpike and onto Holmwood Common, it would have passed to the East of the windmill and to the West of Holmwood House, the residence of Charlotte Larpent, one of the original benefactors of St Mary Magdalene, South Holmwood. The route then followed an alignment that lay slightly to the East of what is now the A24 dual carriageway between South Holmwood and Beare Green, straight through *Bragshaws Farm [sic]* and Palmers Farm, before passing to the West of 'Hullers' [now Hill House Farm] and Misbrooks Green.

To modern eyes, the route of this railway almost seems to have been designed to do as much damage to the ancient buildings, farmsteads and landscapes in this immediate vicinity, with the current day Bregsells Farm [see plan, overleaf] being particularly subject to 'severance and injurious affection' - a phrase explained in the Chapter: A Digression on Compulsory Purchase, see page 17. Had it been built, this line would have been an early example of the unheeding march of progress, seemingly blind to its impact of established communities and driven solely by expediency and the powers of compulsory purchase. As it turned out, it was not until the late 1960s that the full force of this phenomenon would make itself known in the neighbourhood, following the construction of the dual carriageway for the Holmwood 'by-pass'. But that is quite another story altogether.

As an aside, Brunel's perhaps more famous atmospheric railway in South Devon did not open until 13th September 1847. However, when the South Devon Railway Company's atmospheric railway ceased operations on 5th September 1848, it was the end for the commercial and public use of this system of traction.

Meanwhile, during the 1844-45 parliamentary session, another railway line and route entirely was being put forward by the proponents of the London & Portsmouth Railway. This was to run through Dorking, Horsham and Arundel, with branches to Shoreham, Reigate and Fareham.

After leaving Dorking, its alignment was through 'The Punchbowl' public house [demolishing it in the process] and into a tunnel under the Greensand ridge. Emerging from this tunnel, the railway would

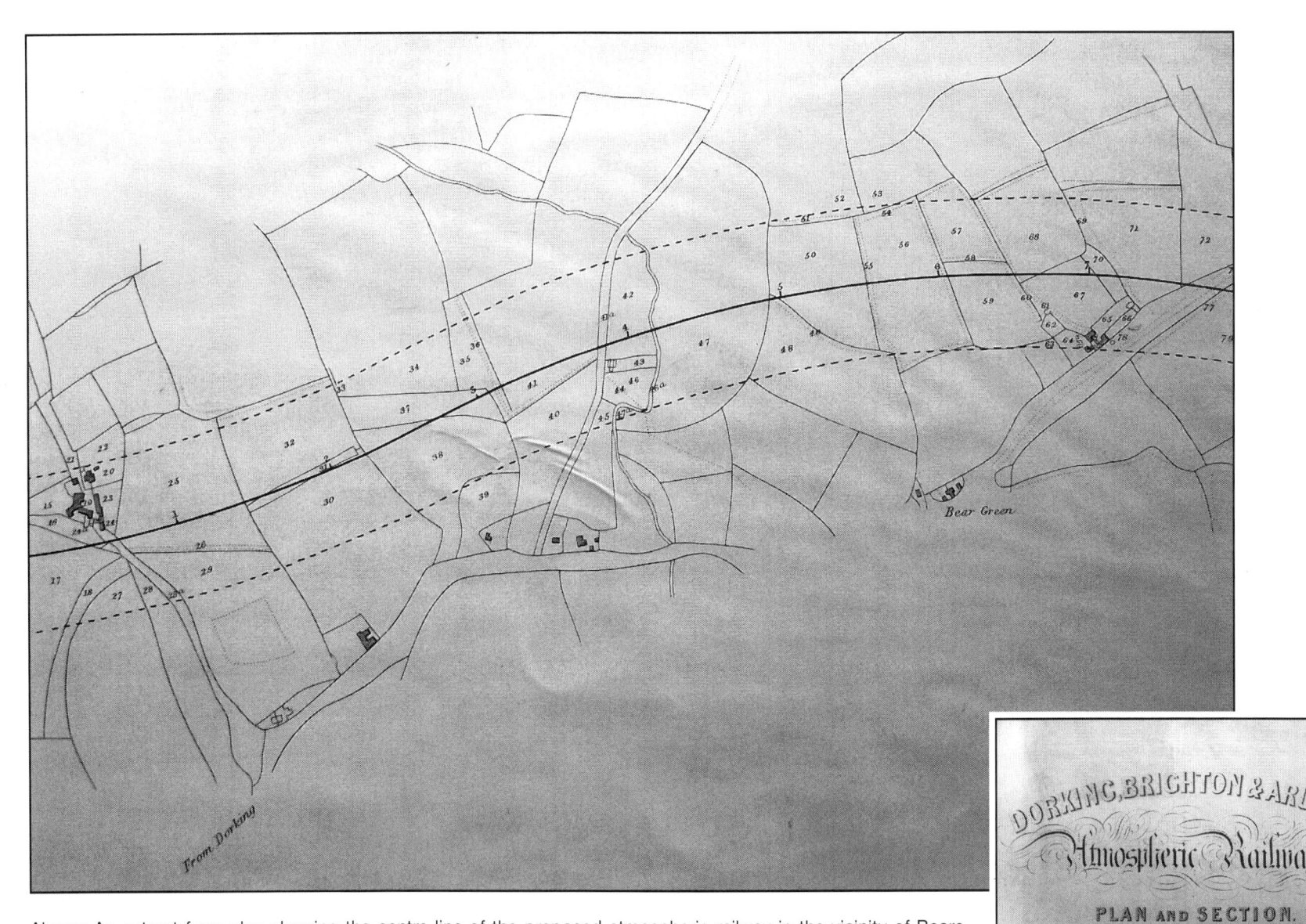

Above: An extract from plan showing the centre line of the proposed atmospheric railway in the vicinity of Beare Green. The line would have run approximately North-South at this point and the modern Bregsell's Farm is to the left of the drawing.

Right: Cover sheet of the route plan bundle for the Dorking, Brighton and Arundel Atmospheric Railway 1845.

have then gone past Root Hill Copse, across Scammells Farm and through Reffolds Copse before passing to the East of Newdigate church and crossing the lane between Dean House Farm and Horsielands Farm. As things turned out, the Bill for this line did not pass through Parliament either.

Nothing daunted, this was not the end of the matter and a broadly similar route in this locality was proposed in 1857-58 for the Shoreham, Horsham and Dorking Railway. Happily, this line managed to avoid 'The Punchbowl', although its projected course did come worryingly close to the buildings at Castle Mill in Pixham. However, like the London & Portsmouth Railway, it would have almost certainly also provided a railway station for the then tiny village of Newdigate [see plan at Appendix: 5]. But the line was not without its critics, as this piece makes clear:

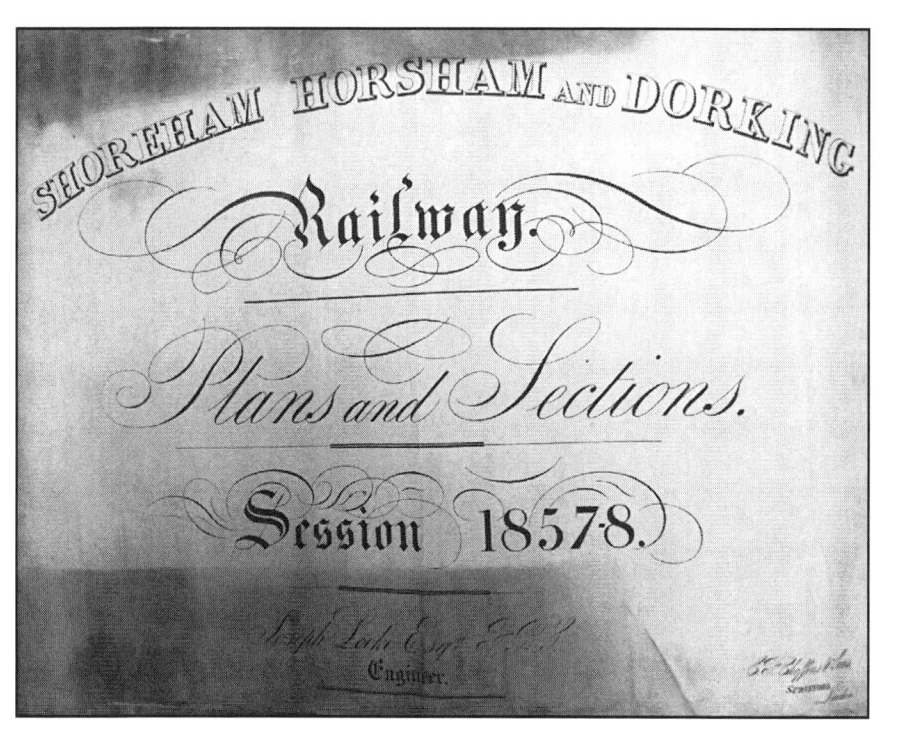

> **The Sussex Express, Surrey Standard, Weald of Kent Mail, Hants and County Advertiser, Saturday 28th November 1857.**
> *DORKING HORSHAM, AND SHOREHAM RAILWAY –*
> For some weeks past a staff of engineers and surveyors have been actively engaged in our neighbourhood, and along the whole route of the projected line, in making the necessary plans and surveys which are now just completed. It appears that a junction with the South-Eastern Railway is to be formed at or near the Box-hill-station. The line will then skirt Pixham-lane, pass close to Messrs Frank's mill, through Betchworth-park, Park-farm, Scammel's-farm, to Brockham-hurst, thence to the village of Newdigate to the east of the church. From Newdigate the rail will run through Taylor's-farm to the parish of Capel, crossing the turnpike road to Warnham, and thence to Horsham, whence we will not follow its course further. It will thus be seen that a very thinly populated and consequently unremunerative district from Dorking to Horsham has been selected for the new line, and we are at a loss to conceive what the intention of its promoters can be in carrying it through such a country. The only village the railway will touch between the two above mentioned towns being Newdigate, the inhabitants of which are but few in number. The population of the Holmwood, Cold Harbour, Leith Hill, Ockley and even Capel, will not receive the accommodation to which they are entitled from the wealth and respectability of each locality. Ockley, we understand, is not within 4¼ miles of the projected railway, and the village of Capel must be at least 1½ miles distant from the nearest point. Altogether, as far as the travelling public are concerned, the new line is not likely to be popular, and we anticipate for its considerable opposition, both on the part of the existing companies and the inhabitants of the districts to which we have adverted.

Left: The cover sheet of the route plan bundle for the Shoreham, Horsham and Dorking Railway.

Overleaf: An extract from the plan showing the proposed junction with the Reading, Guildford & Reigate Railway [which runs East-West] and the route proposed for the Shoreham, Horsham and Dorking Railway, by tunnel through the Greensand ridge.

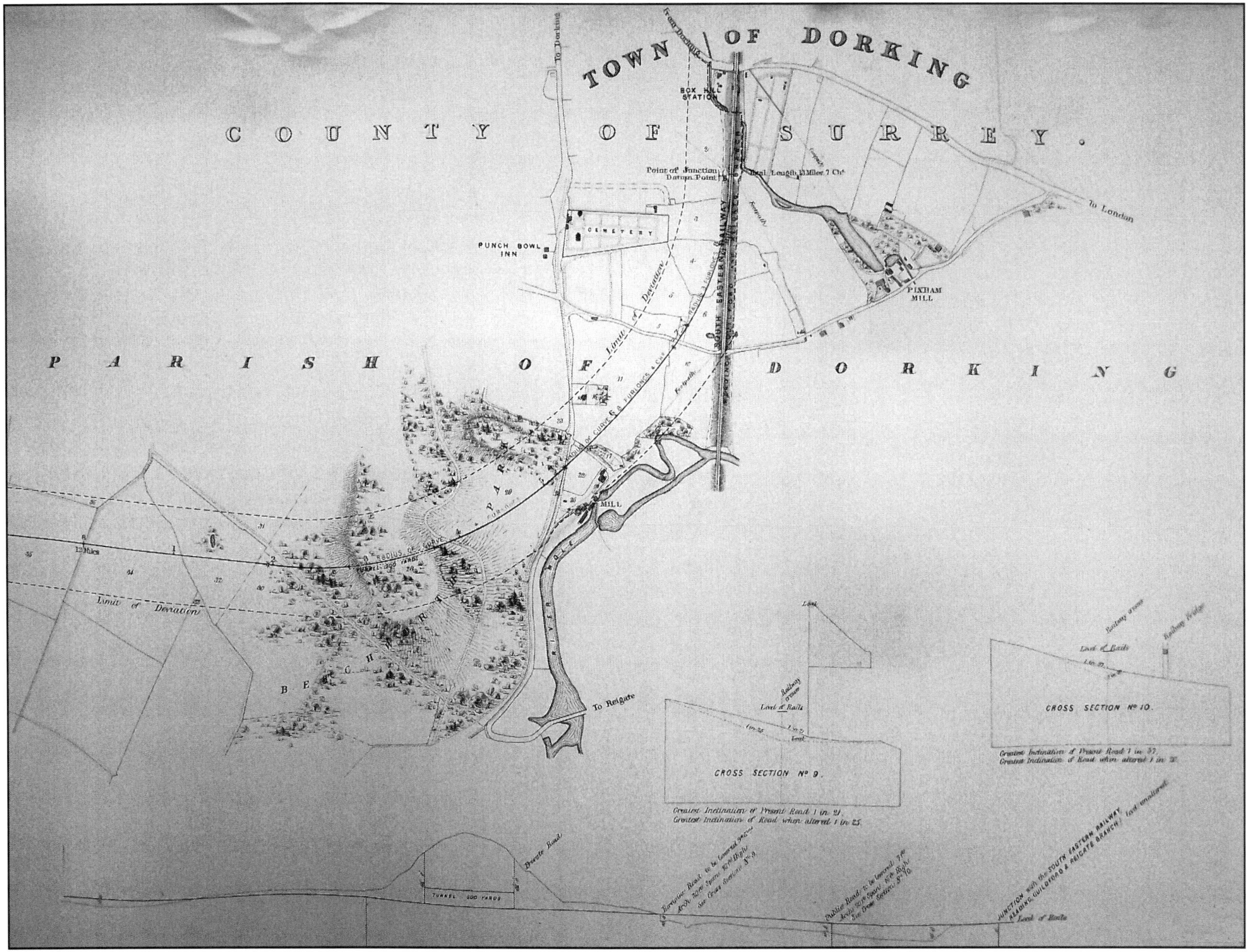
TOWN OF DORKING
COUNTY OF SURREY.
PARISH OF DORKING
BOX HILL STATION
Point of Junction
Datum Point
CEMETERY
PUNCH BOWL INN
PIXHAM MILL
To London
To Dorking
MILL
To Reigate
RIVER MOLE
SOUTH EASTERN RAILWAY
Limit of Deviation
CROSS SECTION No 9.
Greatest Inclination of Present Road 1 in 21.
Greatest Inclination of Road when altered 1 in 25.
CROSS SECTION No 10.
JUNCTION with the SOUTH EASTERN RAILWAY
READING, GUILDFORD & REIGATE BRANCH

Chapter 3
THE HORSHAM, DORKING & LEATHERHEAD RAILWAY

Finally, in the parliamentary session 1861-62, a plan was brought forward for the construction of the Horsham, Dorking and Leatherhead Railway. Amongst several other similar notices placed by the promoters of other proposed railways, the 'Sussex Express' of Saturday 30th November 1861 carried a lengthy notice buried in the middle of page 8 setting out the intention of the Horsham, Dorking and Leatherhead Railway Company to apply to Parliament for an Act that would allow such a railway to be built, "with all proper works and conveniences connected therewith". This notice also described the route and many other legal or technical details, including the intention to apply for compulsory purchase powers. Its full text is set out as Appendix: 2.

Brighton Gazette, Thursday, 13th November 1862
The prospectus has just been issued of the Horsham, Dorking and Leatherhead Railway Company, which was sanctioned by Parliament in the past session, the application for the line from Dorking to Leatherhead having been postponed until the next session, when it is to be made by the Brighton Railway Company. The line authorised from Horsham to Dorking will be 13 miles in length, and will connect the Brighton and South-Eastern lines, besides forming part of a through line of railway communication from London, through Epsom, Leatherhead and Dorking, to the South-Coast and West Sussex lines of the Brighton Company, which converge on Shoreham and Horsham. The capital is £120,000 in £10 shares, and the Brighton Company is to pay a rent equivalent to 4 per cent per annum in perpetuity. The Chairman, Deputy-Chairman and one Director of the Brighton Company are on the Board.

The Bill successfully passed through parliament, receiving Royal assent on 17th July 1862. The Horsham, Dorking and Leatherhead Railway was essentially local in nature and several directors of the company were well-known in the district, including Ockley residents, Mr Lee-Steere from Jayes Park and Mr Labouchere of Broome Hall.

However, money was tight and a deal was eventually brokered with the London, Brighton and South Coast Railway that would allow the construction of double track railway to main line specifications, thereby opening up an alternative route to Portsmouth for the LB&SCR. Owing to these financial constraints, work was only able to start at the southern end of the line in the spring of 1863.

In the end, the Horsham Dorking & Leatherhead Railway was only to build the line from Horsham to Dorking. It was the LB&SCR that was left to construct the remainder of the railway from Leatherhead to Dorking, using powers granted by the LB&SCR (Additional Powers) Act 1864.

The Hampshire Chronicle, Saturday 2nd May 1863
SUSSEX, SURREY, &C.
The first sod of the Horsham, Dorking, and Leatherhead Railway was turned on Saturday, at Horsham, by Mr. Seymour Fitzgerald, M.P. A large number of members of Parliament and gentlemen interested in the undertaking were present. The line is intended to form another route to Brighton.

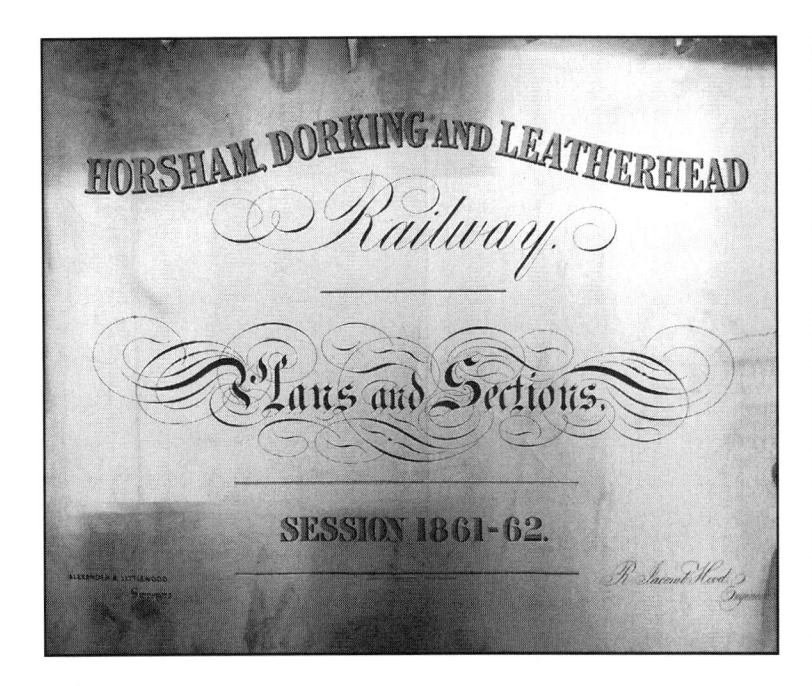

HORSHAM, DORKING AND LEATHERHEAD Railway

Plans and Sections.

SESSION 1861-62.

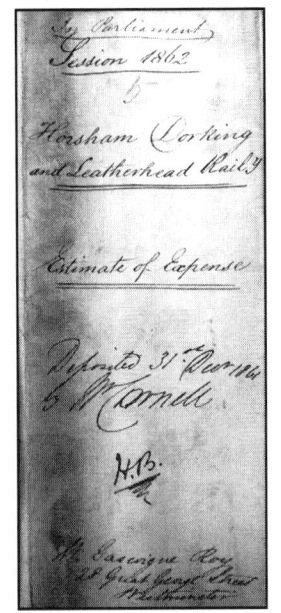

In Parliament Session 1862

Horsham Dorking and Leatherhead Railway

Estimate of Expense

Deposited 31st Decr 1861

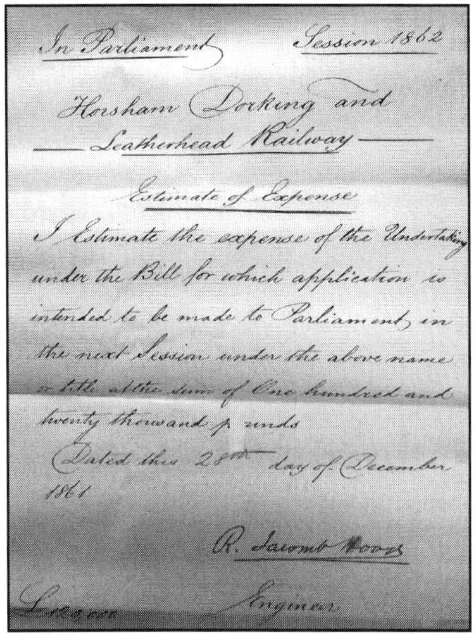

In Parliament   Session 1862

Horsham Dorking and
Leatherhead Railway

Estimate of Expense

I Estimate the expense of the Undertaking under the Bill for which application is intended to be made to Parliament, in the next Session under the above name at the sum of One hundred and twenty thousand pounds.

Dated this 28th day of December 1861

R. Jacomb Hood
Engineer

Chapter 4
TRIALS AND TRIBULATIONS DURING CONSTRUCTION

The construction of the Horsham, Dorking and Leatherhead Railway was not without its problems, some rather more serious than others. In a world where the concept of Health & Safely in the workplace was barely in its infancy, the local press carried far too many articles describing some quite appalling, and often probably avoidable, accidents that befell those working on the new line:

The County Herald, Saturday 4th February 1865
INQUEST.- An Inquest was held on Wednesday last, before C.J.Woods Esq., Coroner, and a respectable Jury, on the body of Elijah Scott, aged 30, who died at the hospital, South Street, on the previous Sunday. It may be remembered that the deceased met with an accident whilst employed in the tunnel now being constructed on the Dorking and Horsham railway, near this town, some eight weeks since, and which necessitated the amputation of one leg at the time. This operation was skilfully performed by Mr.George Curtis, surgeon, at the hospital, where the unfortunate man was taken immediately after the accident, and where, up to the period of his death, he received every possible attention and kindness both from the medical attendants and general staff, and particularly from the Rev.W.H.Joyce, the vicar and Mrs.Joyce. The poor fellow, however, lingered in a hopeless state up to the 22nd inst., when death released him from his sufferings. After the hearing the evidence in accordance with these facts the Jury returned a verdict that the deceased died from the effects of the accident in question.

The County Herald, Saturday 1st July 1865
FATAL ACCIDENT. – On Monday the 19th inst, a fatal accident occurred to a youth named Tidy, who was employed in a cutting on the Dorking and Horsham Railway. It appears that a fall of earth took place, and the poor fellow was partially buried. He was released from his uncomfortable position by his fellow workmen, but he appeared to be in great pain, and a surgeon was sent for. At first it was thought that he was not severely injured, but it was afterwards discovered that a knife, which he carried in one of his trousers' pockets, had penetrated the lower portion of his person, and inflicted a serious wound. In spite of every attention he died in about three hours after the accident had occurred. On Wednesday, Mr.J.C. Woods, the coroner for the western division of the county, held an inquest on the body, when a verdict of Accidental Death was returned.

The Sussex Advertiser, Tuesday 5th December 1865
DORKING - THE ADJOURNED INQUEST ON JONATHAN SHERLOCK, labourer, who had been employed in the construction of the Dorking and Leatherhead Railway, and who was killed by the fall of some soil in a cutting, was held on Monday last. The inquest had been adjourned for the attendance of Mr Sharpe, one of the contractors; but through some defect in the service of the summons, he did not attend. The inquiry was further adjourned, in order to ascertain if the necessary steps were taken to protect the men from injury in the works, and that two of the Messrs. Sharpe might be summoned and their attendance secured.

Far Left Top: The cover sheet of the route plan bundle for the Horsham, Dorking and Leatherhead Railway.
Left: The route of the HD&LR line from Dorking southwards towards Horsham. It skirts Holmwood Common to the East and the village of Capel to the West.
Far Left Bottom: The formal 'Estimate of Expense' for the Line prepared by the Engineer, R. Jacomb Hood. He put the total cost at £120,000.

The Surrey Advertiser, Monday 9th July 1866
FATAL ACCIDENT. – On Thursday last a shocking accident happened to a navvy named Walker, on the new line of railway, and within a short distance of the Dorking railway station now in course of erection. It appears that the unfortunate man was standing on a truck of chalk, which was destined for another part of the line, and whilst the ballast train was in motion he slipped off, and fell under the wheels of the truck. The body was so dreadfully mutilated as to cause instantaneous death.

The Surrey Advertiser, Saturday 10th November 1866
ACCIDENT. – An accident occurred on Friday, the 2nd inst., on the Horsham, Dorking, and Leatherhead Line now in the course of construction, to a labourer named Mark Jackson, who was employed as a breaksman [sic] on the ballast train. It appeared that on the night in question he was proceeding with the train conveying sand from the Betchworth Park Tunnels [sic] to Mickleham, and when near Westhumble, the engine driver slackened speed for the purpose of allowing Jackson, who was then upon the engine (which was pushing the trucks), to step from truck to truck for the purpose of showing his light and ascertaining whether the points in that place were set right to enable them to pass on the proper line of rails, when by some means the unfortunate man slipped down between two, his leg catching between the bumpers [sic], causing a severe fracture. He was immediately conveyed to the Dorking Union Workhouse, and placed under the care of Mr. C. Chaldecott [sic], surgeon, who, on Saturday afternoon, found it necessary to have the limb amputated.

The Surrey Advertiser, Saturday 19th January 1867
ACCIDENT. – A serious accident happened to a navvy working on the Dorking and Leatherhead line on Tuesday last. It appears that the poor fellow, whilst attempting to cross the rail to avoid being run over by the engine, suddenly slipped, and could not make his escape in time to prevent a collision. His arm was completely severed from his body, and he sustained other serious injuries. He was promptly conveyed to the Dorking Union, where he will receive the unremitting attention of Mr. C. Caldecott, the medical officer. The injuries are of such a nature as to leave but little hopes of his recovery.

As one might imagine, the navvies could also be expected to live up to their general reputation for lawlessness from time to time. Yet, on the whole, the district seems to have got off reasonably lightly during the works. At least, unlike other areas in the country, thankfully no riots were reported to have broken out and the newspapers of the time describe only relatively minor offences, more often than not drink related:

The Sussex Advertiser (People's Edition), Wednesday 30th November 1864
CHARGE OF ROBBERY ON THE HORSHAM AND DORKING RAILWAY
William Simmonds, a workman on the line, pleaded guilty to stealing a slab of wood value 6s., from the works of the Horsham and Dorking Railway. – P.C. Albert Standen, who took the prisoner in custody, with the slab of wood in his possession, explained the circumstances of the case, and the prisoner was remanded to the police station until Saturday, Dec 3rd, when, by the consent of the prisoner, the magistrates will dispose of the case.

The Surrey Advertiser, Saturday 25th August 1866
BOROUGH BENCH - MONDAY
(Before the MAYOR and Messrs. WEALE, and TOPHAM.)
STOPPING ON THE JOURNEY – George Nutley, a young man, was charged with being drunk and creating a disturbance at the railway station. P.C. Playford: At five minutes to six on Saturday evening, I was on duty at the railway station. Prisoner was making a disturbance on the platform, and endeavouring to enter a second class carriage. I told him he could not go by the train in the state he was. I prevented him from entering the carriage and told him he had better go away, and come back when he was sober. He would not go away and I took him into custody. He threw himself down and Acting-Sergeant Titley assisted me in getting him to the station. Prisoner in reply to the bench said he had been working on the Horsham and Dorking railway and was going to Winchester. Prisoner was fined 5s.

In some instances, the local press were not averse to creating a storm in a tea-cup over relatively minor, short-lived inconveniences, whilst the long-running uncertainty over the opening date was in danger of becoming a music hall farce:

The Surrey Advertiser Saturday, 11th November 1865

THE HIGHWAYS AND THE NEW RAILROAD – The construction of the new Horsham, Dorking and Leatherhead railway is going on rapidly, but in consequence of the building of a bridge across the highway near Bradley Farm, considerable inconvenience has been occasioned to the public who have to pass that way. From some cause or other, with which we are not acquainted, a very large pool of water has accumulated under the bridge, and the road immediately adjoining is in consequence rendered very soft, and one would be actually up to his knees in the mud there. Horses, vehicles, and people have, of course, to pass this road, and the two former are repeatedly dragging the mud after them, so that it is now quite impossible for foot-people to walk that part of the road – they are obliged, in order to secure a passage, to go over the bridge, and on wet days, which are now not unfrequent, the path over the bridge is not very comfortable, as it is extremely narrow, slippery, and hard-to-be-climbed. If a person falls from it he is certain to be precipitated into the large pool on the road. Complaints are being made from many quarters, and at the petty sessions on Saturday last, Mr T Grissell, one of the presiding justices, spoke in strong terms of the obstruction caused to that part of the highway, by the men building there; and he said there should be something done to try to have it remedied.

The Surrey Advertiser Saturday, 17th March 1866

IS IT NECESSARY? – The works of the Dorking and Horsham railway are progressing. Between Holmwood and Leatherhead several ballast engines are employed daily, Sunday not excepted. It may be fairly asked whether completion of the line is so urgent that the Sabbath, as a day of rest, should not be recognised.

The Surrey Advertiser, Saturday 3rd March 1866

THE NEW RAILWAY. – We understand that many of the people of this neighbourhood complain of the slowness with which the construction of the Horsham, Dorking, and Leatherhead Railway is proceeding. Latterly, however, the completion of the line seems to be getting on brisker than before, and 'ere long we may hope to see it in good working order. No doubt the contractors have an arduous task to accomplish, and, therefore, while it is well for them to push on with the work as speedily as possible, they must, also, of course, keep themselves on safe ground, and not allow their anxiety for the completion of the line to interfere with their caution in having the business well done. The sites for many of the stations along the railway have long since been chosen, and, it seems, are now about to be placed in the hands of efficient contractors. Mr William Shearburn, of Dorking, has already received the contract for the building of the railway stations at Dorking, Mickleham and West Humble, and we believe that no better choice could have been made, as it is well known that Mr. Shearburn fulfils all his contracts with the greatest satisfaction to every party engaged. We understand that the price for the work is not to exceed £12,000.

The Surrey Advertiser Saturday, 12th May 1866

HORSHAM AND LEATHERHEAD RAILWAY. – The question is frequently asked in public assemblies, when will the new line be opened, but strange to say no definite or satisfactory answer can be obtained. It is positively asserted that the time specified in the contract for opening the line for general traffic is early in August next. Workmen are actively engaged both day and night, and judging from the rapidity with which the intermediate stations are being built, it is apparent that the best exertions are being used to complete the whole by the time appointed. The Dorking station is not so advanced towards completion as Leatherhead, but if the weather should prove favourable, judging from present appearances, it will be finished in the course of a few weeks.

A sure sign of progress for any civil engineering project is when the contractor starts to sell off surplus plant – or, as here, horsepower:

The County Herald, Saturday, 30th September 1865
TO CONTRACTORS, BUILDERS, BREWERS, AND OTHERS - SALE OF RAILWAY CART HORSES, DORKING.
BY MESSERS. WHITE & SONS,
Under instructions from Messrs. R. Sharpe and Sons, at the Punch Bowl Inn, near the Box Hill station, on Wednesday, October 4, at One, TWENTY capital Cart or Van HORSES, in consequence of the near completion of the heavy earth-works on the Horsham, Dorking and Leatherhead Railway.

The Sussex Advertiser, Tuesday 3rd July 1866
HORSHAM, DORKING, AND LEATHERHEAD RAILWAY – Notice of Sale of the third portion of the valuable Cart Horses, in consequence of the near completion of the above undertaking.
Messrs. White & Sons
Are favoured with instructions from Messrs R. Sharpe and Sons, the contractors, TO SELL BY AUCTION on THURSDAY, July 12th, 1866, at three for four o'clock precisely, at the Punch Bowl Inn, Reigate Road, Dorking, 30 first rate CART MAREs and GELDINGS, and three NAG HORSES well deserving the attention of contractors, builders, town Carmen, traders, and agriculturalists requiring horses possessing symmetry, strength and activity.
To be viewed on the day previous to and on the morning of sale, at the Stables, near the Punch Bowl Inn, a short distance from Box-hill Station.

But when would the new railway line actually be open to traffic?

The Surrey Advertiser Saturday, 15th December 1866
THE NEW LINE OF RAILWAY. – It is currently reported that the Horsham and Dorking Railway will be opened for passenger and goods traffic early in next year.

The Sussex Advertiser, Saturday 16th February 1867
LONDON, BRIGHTON, AND SOUTH COAST RAILWAY.
... ... The works upon the Leatherhead and Dorking Railway are nearly completed, and the line will shortly be opened. The company will thus obtain access to a new and beautiful district, from which a large traffic may ultimately be expected. The works upon the line from Dorking to Horsham are also approaching completion, and this line is expected to be opened in the present half-year. This line from its completion, in accordance with the provisions of the company's act of 1864, will be amalgamated with and form part of this company's system, and will complete the second line of this company between London and Brighton. ... ...

The Surrey Advertiser Saturday, 2nd March 1867
HORSHAM AND DORKING RAILWAY. – We hear that this new line of railway will be opened for general traffic between Leatherhead and Dorking on Monday next. The directors have made their last inspection of the line and are satisfied with its safety, and the arrangements are so complete as to render a further delay absolutely unnecessary.

The Sussex Express, Tuesday, 26th March 1867
THE DORKING AND HORSHAM RAILWAY. – The Government Engineer made an official inspection of this line a few days since, and reported everything in a satisfactory condition and ready for opening. At present it is not know positively whether it is decided to open for public traffic on the 1st of April or not, it being alleged by some of the officials that this date has been determined upon, whilst others are confident that the event will not occur before the 1st of May.

Right: Even before the Horsham to Dorking railway was opened to traffic and despite the fact that the name of the local station was yet to be finally resolved, its implied impact on property values is shown in these sale particulars produced by Messrs White & Sons of Dorking, in 1866.
Whilst the estate agents had probably followed the contractor's lead and taken the simple expedient of adopting the name of the nearest habitation, Breakspeare *[sic]* Farm, for the new railway station, the LB&SCR chose the name of the nearest community, Holmwood,. This was a commercially prudent decision.

A SMALL BUT EXCEEDINGLY IMPROVABLE
ESTATE
SITUATED BETWEEN
BEAREGREEN & CAPEL

On the West side of the Main Dorking & Horsham Turnpike Road, on which it commands an Important Frontage of 456-feet,

And near to the New Breakspeare Station on the Horsham, Dorking and Leatherhead Railway.

MESSRS

WHITE AND SONS

Are favored with Instructions from the Owner, with the concurrence of the Mortgagee, to Sell by Auction,

AT THE RED LION HOTEL, DORKING,

On Thursday, August 16, 1866,

AT 2 O'CLOCK, IN ONE LOT,

A SMALL COPYHOLD ESTATE

Held of the Duke of Norfolk's Manor of Dorking-cum-Capel, and within the Parish of Capel,

COMPRISING

AN ANCIENT MESSUAGE

formerly, "THE OLD WOOLPACK," which name it still retains;

It is now occupied as Five Tenements with Gardens attached and an Enclosure of exceedingly Productive Meadow Orchard,

Containing altogether about One Acre and a half of very improvable Ground, occupying a frontage of 456-feet on the Turnpike Road, and a further independent frontage on an excellent Public Highway on the South side of the Premises.

The Property is occupied by Mr. John Wood, and will be sold subject to his Tenancy and to the well-known Customs of the Manor of Dorking.

TO BE VIEWED BY APPLICATION ON THE PREMISES.

Particulars and Conditions of Sale may be obtained at the Inns on the Holmwood, Beare Green, Capel, Ockley, Newdigate, Warnham and Kingsfold; at the Anchor Inn, Horsham; at the Offices of J. D. Down, Esq., Dorking, and of Messrs. White & Sons, Auctioneers, Valuers, Land and Timber Surveyors, Estate and Tithe Agents, Dorking.

(R. J. CLARK, PRINTER, DORKING.)

The Sussex Express, Surrey Standard, Weald of Kent Mail, Hants and County Advertiser Tuesday, 30th April 1867

DORKING. – OPENING OF THE DORKING AND HORSHAM RAILWAY. – This event is now definitely fixed for Wednesday next, May 1st. There will be five through trains each way between London and Brighton daily by the route, all of which will stop at Dorking. A more expeditious service from this town to London Bridge and Victoria will take effect from the above date; the time occupied for the journey henceforth being reduced to 55 minutes.

The LB&SCR was formed by Act of Parliament in 1846. Its coat of arms incorporates those of [clockwise from top] London, the Cinque Ports, Brighton and Portsmouth.

Brighton Gazette, Thursday, 2nd May 1867 – Notice on Page 4
LONDON, BRIGHTON, & SOUTH-COAST RAILWAY
OPENING OF THE NEW ROUTE
BETWEEN
BRIGHTON AND LONDON,
Viâ HORSHAM, DORKING AND LEATHERHEAD,
Avoiding the Tunnels.
SEE TIME TABLES FOR MAY.

GEO. HAWKINS
Traffic Manager.
Brighton Terminus,
May, 1867.

The Morning Advertiser, Friday 3rd May 1867
Notice on the front page of the newspaper:

LONDON, BRIGHTON AND SOUTH-COAST RAILWAY – OPENING of the NEW ROUTE to BRIGHTON. The HORSHAM and DORKING LINE, with Stations at Warnham, Ockley and Holmwood,
IS NOW OPEN
for Passenger and Goods Traffic.
The Trains leaving London-bridge for Dorking at 6.5 and 8.20am, and 2.5 and 4.35pm, run through to Brighton by this route.
The Trains leaving Brighton for Horsham at 6.40 and 8.30am, and 12.10 and 4.57pm, run through to London-bridge by this route.

The Horsham, Dorking and Leatherhead Railway 1867 – The Gradient Profile

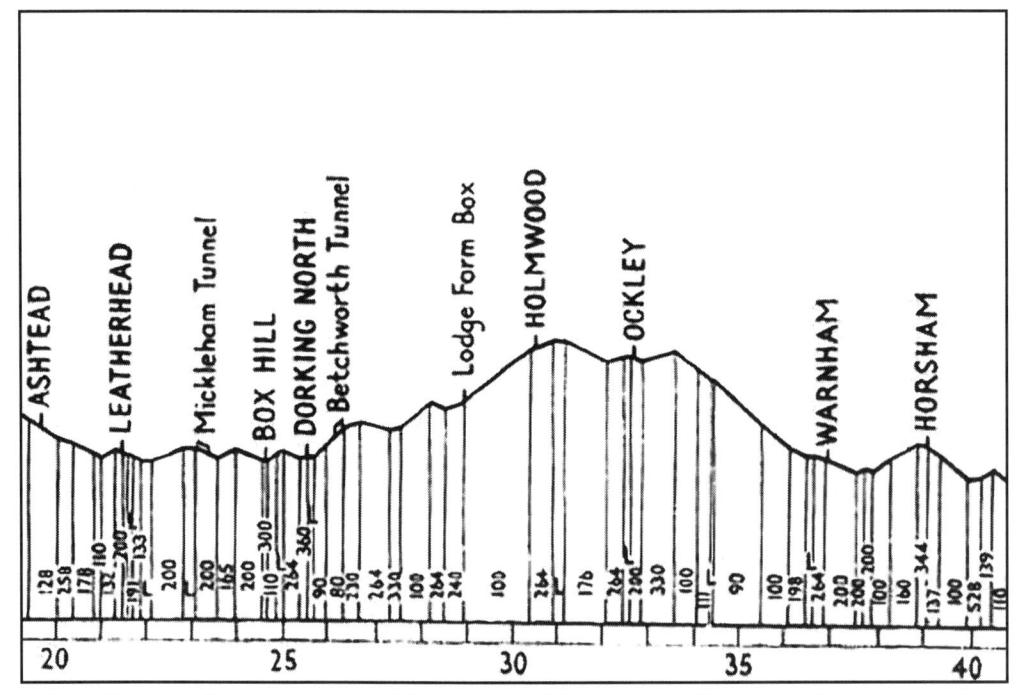

The mile post mileage is measured from London Bridge station. The undulating nature of the line and the climb up to Holmwood station from both Dorking and Horsham is quite apparent.

Chapter 5
A DIGRESSION ON COMPULSORY PURCHASE

During the 'railway mania', that speculative frenzy that took place during the 1840s, the British government promoted an almost total laissez-faire approach to regulation. Once a railway company had obtained its statutory powers, it could do pretty much as it pleased and this freedom created a variety of problems for owners and occupiers of any land required for the construction of the line. With profit being the primary motive, it is perhaps understandable that a railway company would wish to pay as little as possible for the land it needed. Inevitably, such parsimony led to considerable inequity in the compensation sums offered, together with the evolution of some very odd practices, such as only buying part of a house, if that was all that was required, rather than the whole of it.

To help rectify these abuses, Parliament passed the Lands Clauses Consolidation Act 1845, which, amongst other things, ensured that compensation was only determined after professional advice had been provided to the individual owner, occupier or tenant and that the costs of this valuation work and the ensuing negotiations were borne by the promoters of the proposed railway.

The Act also created a statutory process for the parties to follow. In short, the acquiring body was required to issue a notice on those holding an interest in the land to be compulsorily purchased, setting out its intentions, a description of the land in question and an indication that it was "willing to treat for the purchase thereof". In reply to this 'Notice to Treat' (in this context, the word 'treat' means to 'negotiate terms'), the recipient was required to set out their claim for compensation for the land taken and "the damage that may be sustained... by reason of the execution of the works".

Such damage might include 'severance', for example where a field might be split in two or be cut off from its farmstead by the line, or 'injurious affection', ie the damage caused by the presence of the railway to the parts of the holding not acquired. In rural areas, railway companies would often try to mitigate this damage by the provision of 'accommodation works', such as cattle creeps, accommodation crossings or, if needs be, bridges.

As chance would have it, some of the printed ephemera associated with the compulsory acquisition of the land required for some of the railway line to the east of Holmwood and for the station site itself has survived.

The land in question was at *Brackspears Farm [sic]*, a holding owned by Elizabeth Lazenby and in the occupation of a tenant farmer, Richard Charman - see plan overleaf, on page 18. It comprised:

Plot No.	Use	Area		
17	Arable	-	1 Rood	10 Perches
18	Arable	2 Acres	2 Roods	11 Perches
29	Meadow & Stream	1 Acre	0 Roods	11 Perches
30	Meadow & Stream	1 Acre	1 Rood	3 Perches
32 & 33	Arable	1 Acre	2 Roods	23 Perches

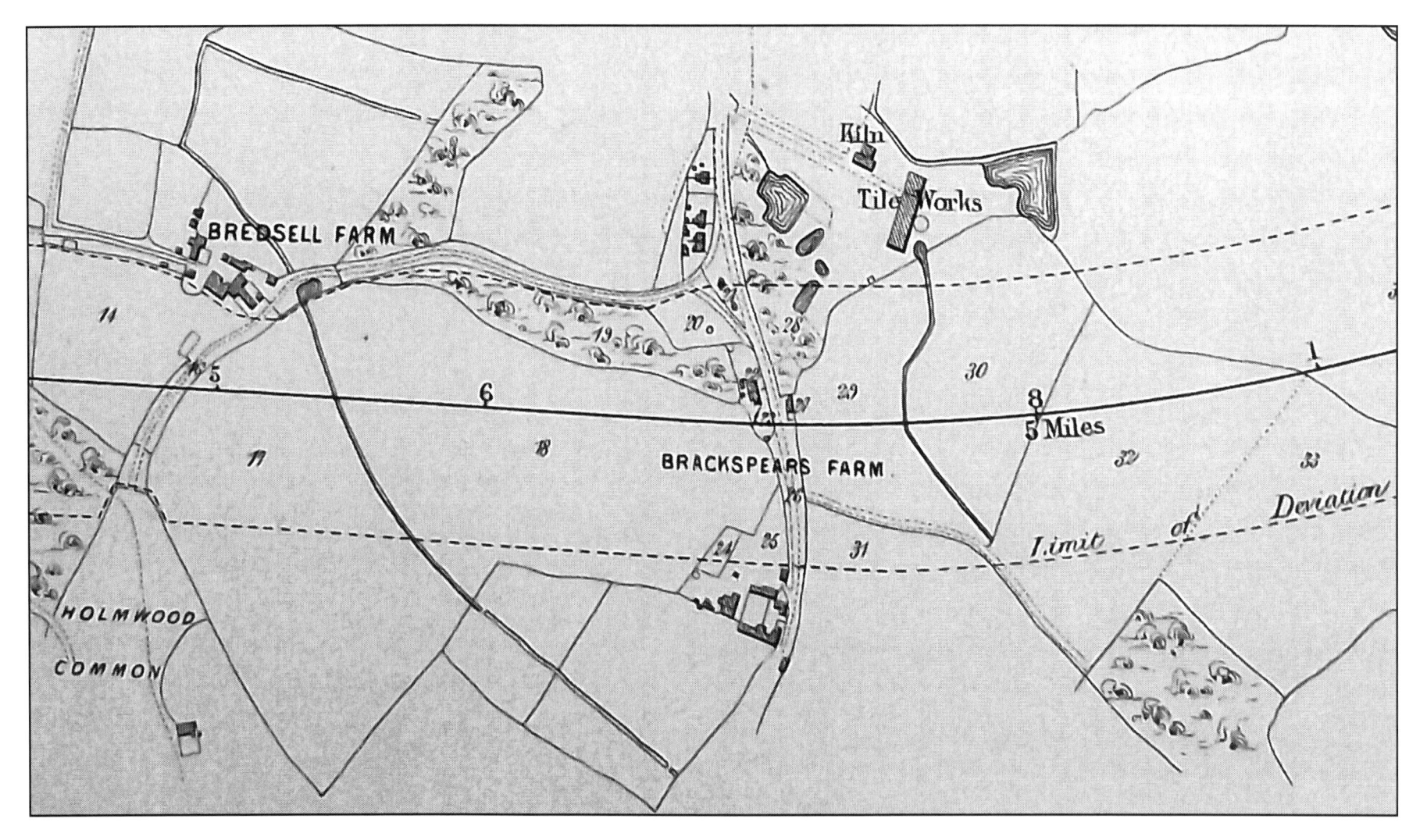

An extract from the Property Ownership Plan prepared for the HD&L Railway, from a point some 4 miles 5 furlongs to 5 miles 1 furlong from Dorking.
The line runs approximately North-East - South-West at this point, ie *Brackspears Farm [sic]* lies broadly South-West of *Bredsell Farm [sic]*.
Apart from the parts of *Brackspears Farm [sic]* set out on page 17, other land needed for the railway at the Holmwood station site included:
Plot Nos. 22 & 23 - A pair of semi-detached cottages owned by John Gilliam Stillwell, occupied by James West Junior & Senior, respectively. These dwellings were demolished to allow the railway cutting to be excavated. A replacement pair of cottages still stand just to the south of the cutting today.
Plot No.27 - A shed owned by the Duke of Norfolk, occupied by James West Snr.
Plot No. 28 - Part of the Brickfield [adjacent to the Tile Works and Kiln shown on the plan] owned by Edward Kerrich, occupied by Jacob Westbrook.

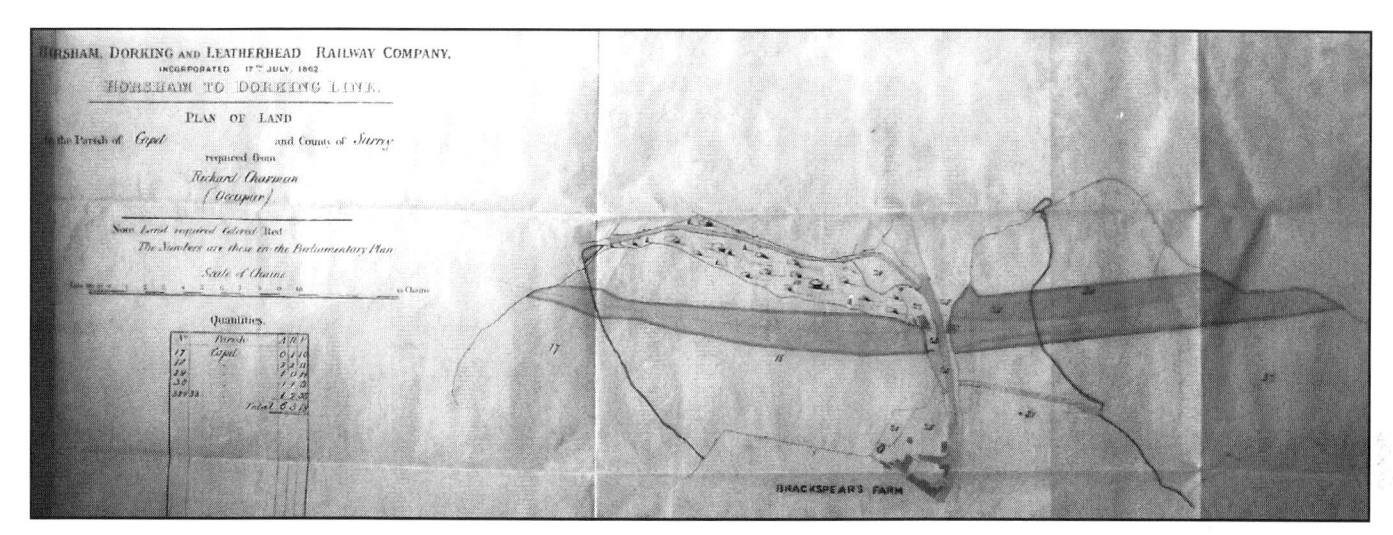

The Notice to Treat for land at *Brackspears Farm* [sic]:

Top: The land acquisition plan.
Left: The front cover of the Notice.
Right: The Schedule, identifying the land, its use, ownership, its occupier and area.
Overleaf The tenant's claim for compensation.

The Schedule of Claim made by Messrs White & Sons in respect of Richard Charman's yearly Michaelmas tenancy is in three parts:

1] The value of his tenant's interest in the 6 acres 3 roods 18 perches of land acquired	£38 1s 6d
2] Compensation for severance in respect of the land retained	£15 11s 10d
3] Compensation for damage done during the initial and subsequent surveys	£ 2 8s 0d
Total	£56 1s 4d

An addendum states that "The crops now on the land are reserved to the tenant".

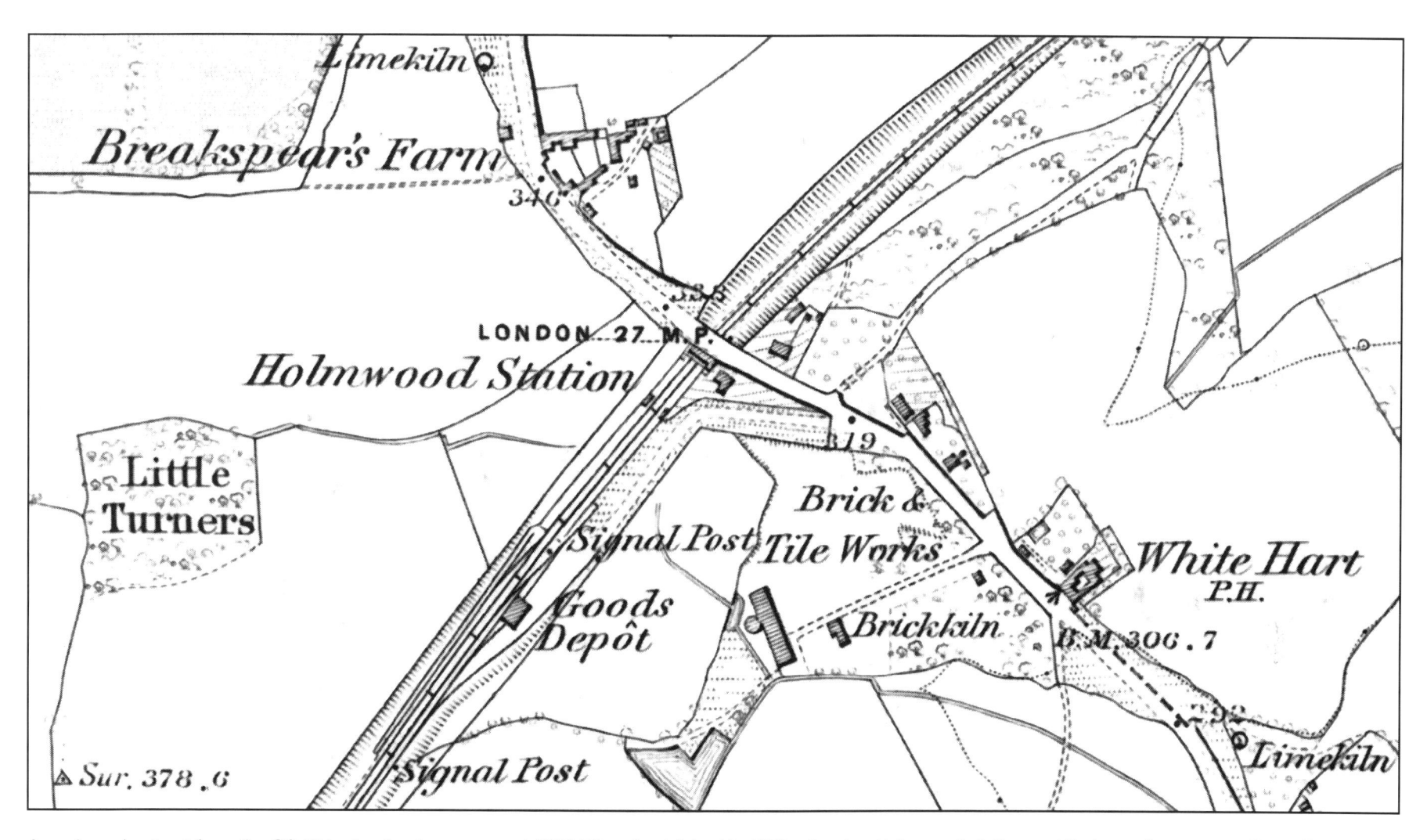

An enlarged extract from the OS 6" to 1 mile plan, surveyed 1869-70 and published in 1873, showing Holmwood station and its immediate surroundings. [This plan is oriented conventionally, with North to the top of the page.] Despite the presence of the lime and brick kilns, or the tile works, the rural nature of the site is immediately apparent. One detail of note is the comparative shortness of the siding to the east of the goods shed, as originally laid. It was soon extended along the boundary fence, towards the station masters house, to provide accommodation for goods trucks serving the staithes occupied by coal merchants. The bay for the cattle dock and the goods shed both date from the opening of the line.

Holmwood Station c.1867 - 1

It is likely that this photograph was taken shortly after the opening of the line in 1867. To the right of the posed group is the Station Master, possibly Benjamin Kerridge, dressed in his frock coat and top hat uniform. Standing to the left is probably his wife, Elizabeth and their three children. Behind her is the Station Master's house, for which a rent of 4/- per week was paid to the LB&SCR, out of a weekly salary of £1-10-0d. The station staff, two porters, probably John Hoad and James West, and the booking & telegraph lad, perhaps Tomas Fairhall, stand between them. The double doors lead into the general waiting room, with the booking office to the left, and the ladies' waiting room, with the parcels & left luggage room behind, to the right. The canopy over the entrance is yet to be added. Everyone, and particularly the photographer, is standing on the turnpike.

Holmwood Station c.1867 - 2

This photograph is pair to that shown on the opposite page and the personae are the same. It shows Holmwood Station at platform level, although in order to take it, the photographer has placed his tripod in the middle of the up line. Behind the Station Master is a platform shelter which lasted until the late 1980s. Its equivalent on the up platform is still standing. Beneath each flight of stairs, in both basements of the station building, there is a WC and on the down side the windows of the lamp room are clearly visible. The up starting signal can just be seen through the bridge opening. To enable the signalman to view this and other signals on the long curve through the station, and operations generally, a squat signal box was built in 1877 on the up platform. The pedestrian bridge at the rear of the station building and the stairways leading down to platform level were originally open to the elements.

Chapter 6
THE EARLY YEARS OF HOLMWOOD STATION

As announced in the press, the new railway and the station at Holmwood were opened for business on Wednesday, 1st May 1867. At that time, the main centre of population was at South Holmwood, well over half a mile to the North of the line, with a smaller group of cottages at Beare Green, about half a mile to the South. Set back from the turnpike or scattered throughout the neighbourhood were small farmsteads, gentlemens' residences or isolated individual dwellings. The station was built where the line crossed the Dorking - Horsham turnpike, adjacent to a brick and tile works. The usual passenger facilities were provided and commercial freight traffic was served by the small goods yard, with its shed, cattle pen and crane.

However, the LB&SCR was in financial difficulties and, in the financial crisis of 1867 that followed the collapse of the bankers Overend, Gurney & Co. in 1866, on the brink of bankruptcy. Large capital projects, such as the HD&LR, had over-extended the Company's reserves. The new works were to have been sustained by profits from newly-generated passenger traffic, but because of the economic downturn this income failed to materialise. Indeed, several of its country lines were losing money and prospects were bleak.

So, with passenger traffic falling short of expectations, the LB&SCR response was swift, whilst being, it seems, as unpopular as it was perhaps predictable[see newspaper cutting opposite].

The sense of frustration expressed in this article is palpable and its sentiment is clearly one that is still recognisable to the passengers of today. Fortunately, thanks to the business acumen of the replacement LB&SCR Chairman, Samuel Laing, and the new Secretary/General Manager, J.P. Knight, the Company returned to a more stable financial footing during the early 1870s.

> **The Sussex Express, Tuesday 8th October 1867**
> LONDON AND BRIGHTON RAILWAY – TWO trains between Dorking and Brighton have been withdrawn this month, viz, the 1st up and the 1st down. The earliest arrival from Brighton now is 9.19a.m. and the earliest departure from that place 9.14a.m. As the former is nearly three hours on the road - it leaves Brighton at 6.35 a.m.- it is perfectly useless for all purposes as far as Dorking is concerned. The result of this strange arrangement is that early all the traffic to and from Brighton and this place takes the old route via Red Hill, and will continue to do so. Practically, the new line might as well be closed altogether.

Although the discovery of the delights of Holmwood, together with its common and the southern slopes of Leith Hill, as charming spots on which to create substantial country residences for wealthy folk pre-dated the coming of the line, the railway slowly started to make its mark. The presence of the newly opened Holmwood station made the location even more desirable for those wishing to live in deepest rural Surrey and still have relatively easy access to the Metropolis.

This point was not lost on the estate agents and auctioneers of the period, as the advertisement on the opposite page makes clear:

> **The Morning Post, Monday 11th October 1869**
>
> *Sale by Auction*
>
> Holmwood, near Dorking, Surrey
> ~ The Oakdean Estate ~
> A most desirable and attractive Freehold Residential Property, 10 minutes' walk from the Holmwood Station, comprising an excellent family residence, with capital stabling and small farm-yard, pleasure ground, shrubberies and plantations, productive kitchen garden, young orchard, and a small grazing farm, the whole about 32 acres, studded with handsome timber trees, and occupying an elevated and commanding position in this much admired and favourite residential locality within easy reach of the City and West-end of London.
> MR. JAMES BEAL has been instructed to OFFER the above for SALE, by AUCTION, at the Mart. Tokenhouse-yard, E.C., on THURSDAY, October 28, at one o'clock.
> Particulars and cards to view may be obtained at the Mart., of Messrs. Tilleard and Co., solicitors, 37, Old Jewry,
> and of the auctioneer, 200, Piccadilly.

Another prime example of the impact of metropolitan money on the neighbourhood was at Kitlands, Coldharbour. The original farm was sold in 1823 by the Bax family to George Heath, a Serjeant-at-Law [in modern parlance, a barrister] practising from chambers in London. Mr Heath gradually extended and improved the house and grounds, transforming it into a gentleman's country residence eminently suitable for his status in life. As the income from his legal practice continued to grow, Moorhurst, Anstie Farm and Trouts Farm were also added to the estate. In 1861-62 George Heath's second son, Admiral Sir Leopold Heath, built Anstie Grange on a prominent site located between Coldharbour and South Holmwood, with extensive views across the Weald.

Admiral Heath was the first of the many people prominent in national affairs to use Holmwood station. He had joined the Royal Navy in 1830 and was involved in the capture of Borneo in 1846, but it was during the Crimean War that he made his name. He was the beach master during the British landings at Eupatoria in 1854 and went on to take personal charge of the Port of Balaclava in 1855. It was prize money from this war that funded the construction of Anstie Grange. Thereafter, in 1863, he was appointed Vice-President of the Ordnance Select Committee at Woolwich; in 1867 he was Commander-in-Chief, East Indies Station; in 1868 he oversaw the naval aspects of the expedition to Abyssinia; in 1870 he served on the committee for torpedo defence before finally retiring from service in the Royal Navy in 1877.

In his retirement, Sir Leopold became a Director of the Hand in Hand Fire & Life Insurance Society [the oldest such Company in England]; of the Central Bank of London and of the Eastern and South African Telegraph Company. When not occupied travelling up to London to carry out these duties, he was also a magistrate for the County of Surrey. He fathered seven children, two of whom went on to become major-generals in the British Army, one an Admiral in the Royal Navy and yet another, Cuthbert, unable to pursue a military career owing to deafness acquired in childhood, was the maker of the modern Lloyd's of London and became a millionaire many times over.

At the other end of the social scale, the new railway line opened up other possibilities, including the boarding-out of urban pauper orphans in rural communities. A lengthy and verbose article in an 1872 edition of the Daily News describes this practice, but so discursive is the piece that one is left with the distinct impression that the author was probably being paid by the word and, in view of its patronising tone and ramblings, that he may also have dined rather too well before it was written. Yet, in spite of these flaws, the piece essentially describes rationale and economics that underpinned the system: simply, it was cheaper than keeping and educating the orphans in the metropolitan Union Workhouses.

ADMIRAL
SIR LEOPOLD GEORGE HEATH, K.C.B.
1873.

The Surrey Hills, Frederick Ernest Green 1915, pages 141-2

At Anstie Grange lived until a few years ago, when he died, Admiral Sir Leopold Heath, who was one of the heroes of the Crimea. Bearing in mind that Leith Hill seems to have in it something which develops eccentricity in elderly gentlemen, I should like to relate a little incident that occurred one day at Holmwood station, in my hearing. The old Admiral drove up in his carriage, and, after giving many minute directions to his coachman a ritual of instructions surviving from navy days and evoking monotonous responses from the coachman of "Yessirleppold" (which ended with a wink to a crony nearby), the fine old sailor was accosted by a young man, who greeted him as a friend.
"I don't think I know you," remarked the old gentleman, who was a little deaf.
"Oh yes, you do," said the young man, full of youthful assurance; "you met me at Lady ---'s."
"Oh, did I ... The Colonial Office, did you say? How do you like Mr.---?" (mentioning the Colonial Secretary).
"Very well. Are you a Free Trader or a Tariff Reformer, Sir Leopold?"
"That's of small moment," replied the old sailor curtly. "What we want is a strong Government and particularly a strong navy. Nothing else matters. You ought to come and listen to me at the village institute; I am going to speak there to-night. Yes, a strong navy ... a strong navy ... that is all we want".
I pitied the self-assured young civil servant, who looked a little crestfallen as the Admiral walked down the steps to meet his train.

To be candid, I'm not entirely sure that I fully understand the point of this anecdote either! JSW

The article describes a visit to Holmwood and manages to incorporate references to not only the methods of teaching at Holmwood school and the domestic arrangements of the rural poor in the district surrounding the station, but also the 1871 novella, the 'Battle of Dorking' [see Chapter 7], and the progression of a performing bear through Beare Green and Holmwood. Despite its length, so interesting is the piece that it is set out in its entirety as Appendix: 3, suitably clarified with the addition of contemporaneous illustrations. However, these brief extracts set the scene in respect of Holmwood station:

The Daily News, Thursday 1st August 1872
BOARDED OUT

… For the last year or two the English guardians have been cautiously trying the system of boarding-out … … There are also children at Holmwood, in the same county, and these the writer went to see the other day, and to report for himself. Holmwood is close to Dorking, and Dorking, as everybody knows, is very nearly the loveliest place in the south of England. The traces of the late dreadful battle have entirely disappeared … … In one of these bye-lanes not far from the Holmwood station, the writer, being under clerical guidance, got on the track of two infant boarders out. The cottage, which was their new home - exchanged for a ward in the Poplar Union - might have been higher and broader with advantage, but it could not have been cleaner, or - what is something to the purpose - more picturesque.

Meanwhile, the exacting nature of work on the railway was beginning to make itself felt and the London Brighton & South Coast Railway was clearly a hard task master. The LB&SCR Traffic Department Registers record the staff employed at Holmwood station and, in addition to the expected promotions, transfers and resignations, they also hint at the strict discipline expected of railway servants. The first members of staff to be dismissed from their posts at Holmwood station were Porter John Foster and Goods Porter James West. Both were discharged on 9th January 1869, Foster for being "Absent without leave" and West for "Intoxication". Whether it was a joint enterprise that led to this outcome is not clear, but it does seem probable that the proximity to the station of the 'White Hart' public house, and the temptations it provided, may well have played a part in this outcome. Interestingly, Foster was reappointed by the LB&SCR at Burgess Hill in March 1869.

The 'White Hart' Hotel, c.1905.

The turnpike leads away from the photographer and up the hill towards the railway station, about 200 yards away. The entrance to the goods yard is just in view, on the left at the bottom of the hill.

In the late 19th century, this hostelry brewed its own beer. The auction particulars produced for a sale of the property in 1880 record that the brewery was approached through the large double doors to the right of the premises and contained a 7-barrel capacity copper. The water supply for this production came from a 170 feet deep well "of excellent water" in the garden at the rear. In the yard, there was a 6-stall stable and piggeries. Behind the cart, stopped in "the pull-up space in front of the house", there was a further stable "conveniently placed for travellers' horses". The original plain, late-Georgian, architectural features of the property are in marked contrast to the alterations made after the Great War.

The next railwayman to be discharged, on 17th March 1869, was the Booking & Telegraph Clerk, Thomas Henry Fairhall, for "not accounting for cash received". This ought to have been a warning to one and all, and especially the Station Master, Benjamin Kerridge, who presumably played a significant part in Fairhall's departure. So it is somewhat surprising to see that Kerridge himself was dismissed on 2nd March 1874 for "Irregularities in cash accounts". Even more amazing still, the replacement Station master, James Bateson, was also discharged on 15th January 1875 for what appears to be a very similar lapse, "Deficiencies in cash accounts".

But it was not to be long before the LB&SCR had identified a suitable replacement for the disgraced Bateson. As this press cutting succinctly reports, Mr Edward Longhurst was, in the parlance of the Registers, "removed" from Emsworth, in Hampshire, to Holmwood. Still at Holmwood, he retired from railway service on 31st December 1896, aged 70.

But perhaps the most unusual misdemeanour at Holmwood during this period was that committed by Booking & Telegraph Clerk, Ernest Bowers. He was dismissed on 25th November 1870 for "Stealing pair of boots" - whether they were Company-issue or goods in transit was not recorded in the Register.

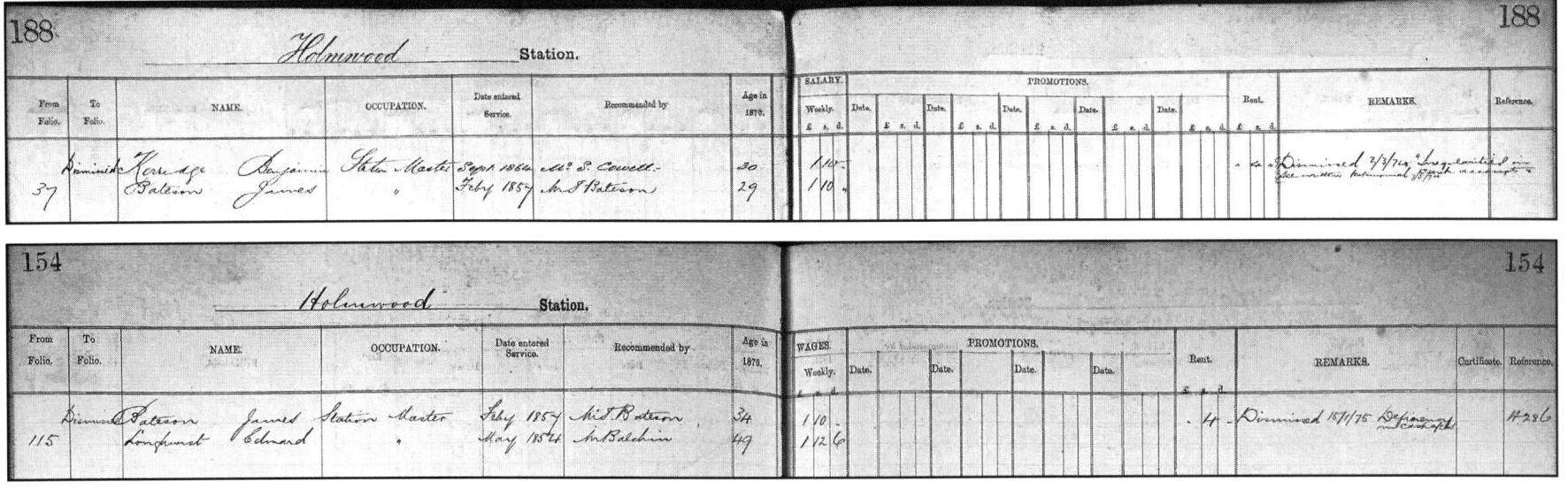

> **The Hampshire Telegraph & Sussex Chronicle,**
> **Saturday 20th February 1875**
> ***EMSWORTH.***
> A testimonial was, last week, presented to Mr. Edward Longhurst, late stationmaster at Emsworth, who has just been promoted to Holmwood Station. Mr Longhurst has been five years at Emsworth, and during that time he has gained the respect of the inhabitants of the town and neighbourhood. As soon as his removal was spoken of, a subscription was started, and £36.12s.was quickly raised and presented, accompanied by a written testimonial, setting forth, in glowing terms, the efficiency, kindness, and willingness to please of the recipient. The testimonial, which bore the names of the leading residents, was presented through Mr. George Clarke, of Emsworth.

Above: Extracts from the LB&SCR Traffic Department Registers for 1870-74 [top] and 1875, showing the fate of Station Masters Kerridge and Bateson

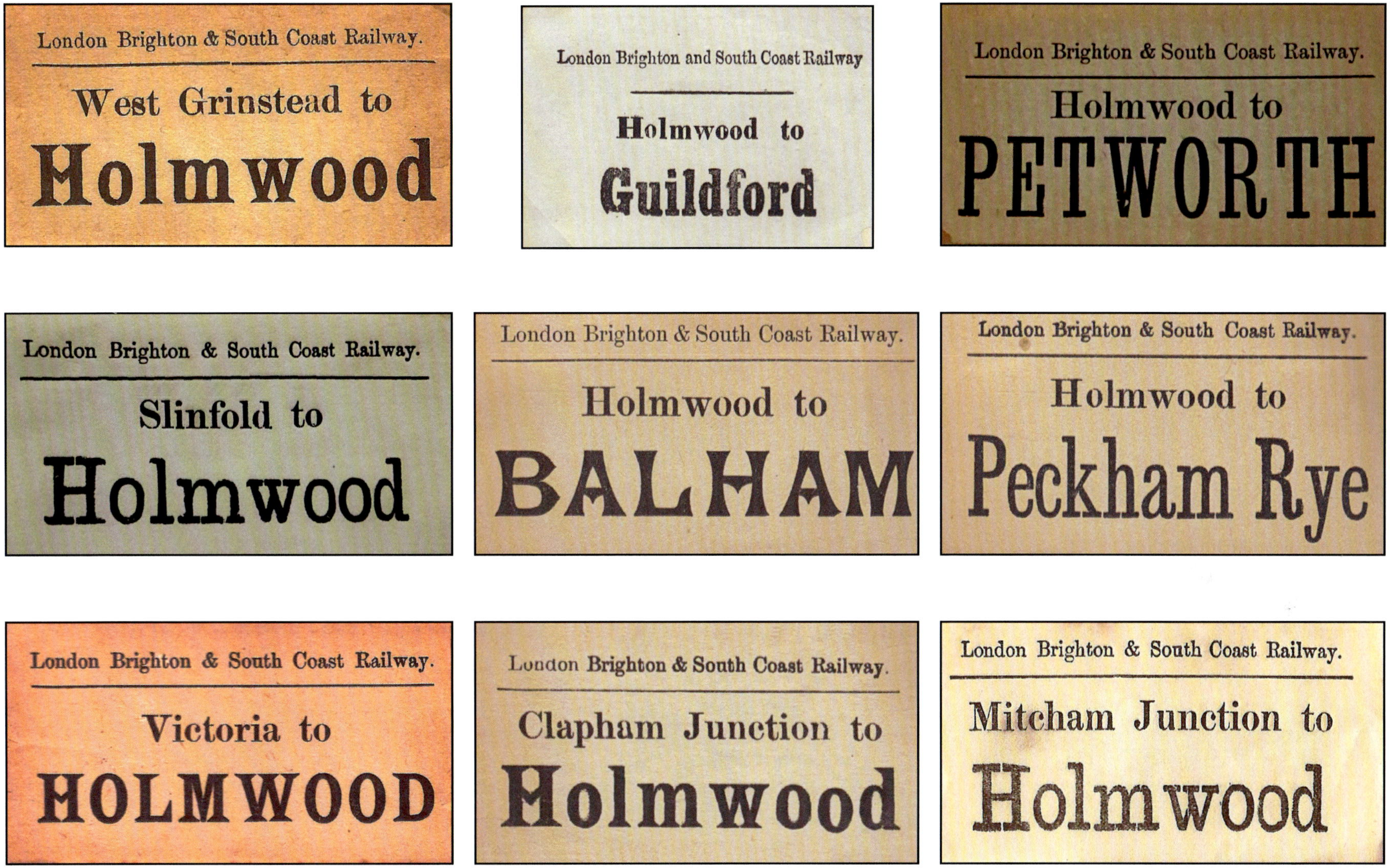

Printed Ephemera – 1

London Brighton & South Coast Railway luggage labels, to and from Holmwood

Chapter 7
'THE BATTLE OF DORKING' AND ITS AFTERMATH

For the avoidance of doubt, 'The Battle of Dorking' is an entirely fictional creation. It was the work of George, later General Sir George, Tomkyns Chesney KCB CSI CIE [1830-1895]. Published, initially anonymously, by Blackwood's Magazine in May 1871, it is a convincing, albeit melodramatic, account of a supposed invasion of England by a German-speaking enemy and the ensuing disintegration of the British Empire. It first appeared as a supplement to the magazine, but proved so popular that it was reprinted as a sixpenny booklet and sold 80,000 copies within the first month. This edition is only 64 pages long, yet the impact of its message on the national psyche and the previously perceived invincibility of the British Empire was profound, following as it did in the wake of the Franco-Prussian War, the Siege of Paris and the civil insurrection during the Paris Commune in the years 1870-71. If the proud citizens of the French capital could be reduced to eating the animals in the Paris zoo and eventually wind up shooting and killing each other on the barricades, in the streets or by summary execution, what might happen in London and the south of England in the aftermath of a similar successful invasion?

Indeed, further verisimilitude is provided in the story by the troop deployments using recognisable railway lines, whilst various skirmishes and the actual battle itself are all set in, and constrained by, well-known geographical features around Dorking.

Early in the tale, a volunteer army sent to fend off the Germanic foe. The narrator tells of being taken by train from Waterloo, vîa Leatherhead, first to Dorking and thence "to a small station" a few miles short of Horsham. Here, "the order came to leave the train, and our brigade formed a column on the highroad".

Some literary commentaries suggest that this station could have been [whisper it quietly] either Ockley or Warnham. But the application of local knowledge, or merely a few moments study of a map, will show that neither of them is situated on a 'highroad'. Whereas, in marked contrast, Holmwood station most definitely is.

However, perhaps the most compelling argument for the strategic significance of Holmwood station came with the British Army's response to the furore that followed the novella's publication. For the rest of the decade, military manoeuvres were undertaken to test and affirm the capability of the armed services to defend against invasion from continental Europe.

But before looking at two newspaper articles that describe this reaction, it is worth remembering that, during the 1870's, anybody walking with a limp or with an arm in a sling may well be asked if they had been wounded at Dorking!

The first describes where and how army exercises and troop deployments in Surrey, Sussex, Hampshire, Wiltshire, Gloucestershire and Somerset are to be made, before going on to describe the provision of compensation payments:

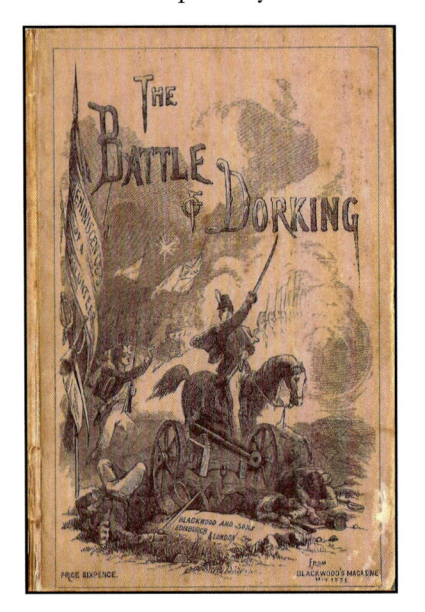

> **The Hampshire Telegraph & Sussex Chronicle**
> **Wednesday 14th June 1876**
> *THE TRAINING OF ARMY CORPS*
> The Secretary of War's Bill "to facilitate the assembling and training of certain corps" was published on Saturday. It provides for the calling out of the Second and Fifth army Corps during the summer, and authorises "the forces, with their arms, munitions of war, and stores, "to pass over, encamp, or practice military exercises on any lands, and use and public, private, or occupation roads", within the limits of the following areas:-
>
> *TRAINING GROUND OF THE SECOND ARMY CORPS.*
> AREA 1. – IN THE COUNTY OF SURREY. – An area bounded by a line starting from the Shalton [sic] junction of the Reading and Redhill branch of the South-Eastern Railway with the Guildford and Portsmouth branch of the London and South-Western Railway, about a mile south of Guildford, and running along the said branch of the South-Eastern Railway in an easterly direction to the point where it crosses the Leatherhead, Dorking, and Horsham branch of the London, Brighton, and South Coats Railway in a southerly direction to Holmwood station; thence along an imaginary straight line in a south-westerly direction to a point where the said straight line cuts the Guildford and Horsham Branch of the London, Brighton, and South Coast railway, near Lower Wipley [sic] Farm; thence in a north-westerly direction along the said branch of the London, Brighton, and South Coast Railway to its junction with the Guildford and Portsmouth branch of the London and South-Western railway at Pease Marsh; and thence in a north-easterly direction along the said branch of the London and South-Western Railway to its junction with the Reading and Redhill branch of the South-eastern Railway at Salford [sic] aforesaid ...

> The bill prescribes the mode of determining compensation payable in respect of damage by the passage of the forces to be operative till the 1st of June next year; but the act, so far as it relates to the power of the forces to occupy land, &c, is not to remain in force after the 1st of September next.

If the 'The Battle of Dorking' was taken literally by its readers, and many seemingly did so in the early 1870's [see Appendix: 3], the spiked helmets of the invader would first have seen in Holmwood in the summer of 1876. In reality, troops from over the water did arrive - but it was actually the Armagh Militia that marched up the turnpike from Holmwood station to their camp in that year, and not the Hun:

> **The Belfast News-Letter, Monday 17th July 1876**
> *THE ARMAGH MILITIA IN THE FIELD*
> *HOLMWOOD CAMP, DORKING, FRIDAY EVENING.*
> I remained here to-day to see the Armagh Light Infantry in camp, leaving until to-morrow to pay a visit to the Irish Brigade at Horsham, when the two corps will put in an appearance, and complete their camp on Saturday. The Armagh and the Monaghan corps steamed into Spithead on board the troopship Orontes on Thursday evening, and were this morning dispatched, the latter to East Codford, to complete the 1st Brigade 2nd Division 5th Army Corps, and the former to Holmwood.
> I was present when the Armagh Light Infantry, under the command of Colonel Cross, marched up to their tents at about one o'clock today. The weather was extremely hot, it being, as I can certify from past as well as present experience, thoroughly tropical, and yet the men, who are very fine fellows, came in apparently in no way distressed after their three days' close confinement on shipboard, their hurried and crowded transit to the Holmwood station, and terrifically hot, albeit short, march in heavy marching order to the camp ...

This article also included an extract from the 'London Standard' which suggested that "a hundred men of the regiment deserted to the mountains the moment they heard they were to embark for England, the idea having got abroad amongst them that 'it was intended to send them for service to the East'". Having set up this defamatory tale, the piece goes on to rather condescendingly denounce it, "Nobody who knew aught of Ireland, or of the Irish militia, can be supposed to have given a moment's credence to this story; but there are innocent people who will believe anything they see in print, and who are prepared to burden the inhabitants of the sister island with ignorance enough to mistake Surrey for Servia".

It then concludes:

Station Hill, Beare Green.

Pub. by W. H. Geary, Beare Green.

The Belfast News-Letter, Monday 17th July 1876
The site of the "Battle of Dorking" is the only scene of warfare past or present, real or imaginary, that the gallant fellows are likely to visit before they are dismissed at the end of the month.

Above: A locally produced post card showing mounted soldiers, probably from the Surrey Yeomanry, outside Holmwood station, c.1900-1910. Note the entrance to the goods yard to the left of the picture and the lane down to Bregsells Farm on the right, between the white-painted picket fences.

Chapter 8

BETCHWORTH TUNNEL:
ITS COLLAPSE & OTHER MISADVENTURES

The Betchworth tunnel, actually under Deepdene Park rather than being in Betchworth village, is the most significant civil engineering feature on the railway between Horsham and Dorking. It is straight for the whole of its 385-yard length and on a rising gradient of 1 in 80 towards Holmwood. Although it had only been open to traffic for just over twenty years, it suddenly collapsed at its northern end on 27th July 1887. A calamity was avoided only because of sheer good fortune and a clear-thinking railway ganger. Alas, the various newspaper reports of the incident and its aftermath were not always entirely accurate, hence the addition of *[sic]* against various inconsistencies.

The local paper published this condescending and infuriatingly patronising whinge some four months after the event, which unfortunately does absolutely nothing to ameliorate the position or inform its readers in any sensible way:

> **Dorking Advertiser, Saturday 26th November 1887**
> The collapse of the tunnel, a little distance from Dorking, appears to have caused a great deal of inconvenience to those who are in the habit of travelling towards Horsham. Betchworth tunnel, lying as it does between Holmwood and Dorking occupies a very strategic position, and its temporary blockade has been a source of much confusion. We are glad to learn that the railway authorities have made some alterations in their system of trains, which will confer a boon on the public. We heard a good deal of grumbling at Dorking the other day by would be travellers to Horsham, so that the step taken by the railway officials does not come a bit too soon.

More helpfully, other newspapers provided a rather better insight into the causes of the collapse and its aftermath. But things would take a turn for the worse before it became possible to start trains running through from Dorking to Holmwood again:

> **Lloyd's Weekly London Newspaper, Sunday 31st July 1887**
> *LANDSLIP IN A TUNNEL.*
> On Wednesday afternoon a train had a narrow escape from being buried by a landslip in the Betchworth tunnel. Just outside Dorking station, on the London and Brighton company's main Portsmouth line, is a very deep cutting, leading into Betchworth tunnel, which is about a quarter of a mile in length. The 4.25 up train from Horsham ran safely through on Wednesday afternoon, but a few seconds later a ganger who was walking through was alarmed by some bricks falling about him. He at once rushed out, and tons of earth fell with a great crash, carrying with it the roof of the tunnel for many yards. It is supposed that the heavy traffic resulting from the Goodwood races and the naval review led to the slip, as many heavily laden express trains had run through the tunnel during the last fortnight. The slip occurred near the spot where several *[sic]* men were killed by a fall of earth when the tunnel was being constructed. The traffic was entirely suspended, the passengers by the 24 minutes past six Portsmouth up *[sic]* express having to leave the carriages and walk up the high embankment *[sic]* [into the park, their luggage being brought by the same route. The slip viewed from the top presented the appearance of a deep chasm nearly 100 yards in circumference. On Thursday the preliminary steps were taken for clearing the tunnel, but it is believed that some weeks must elapse before the line will be again opened for traffic.

Dorking & Leatherhead Advertiser
Saturday 10th December 1887
BETCHWORTH TUNNEL.

The chief engineer of the London, Brighton, and South Coast Railway Company visited the works in progress at this tunnel last week, and found that the older portion thereof was also in a dangerous condition. As a result of this examination the company determined to re-line the tunnel throughout, and Messrs. Firbank, the present contractors, have received an order to push the extra work to as speedy a termination as possible. It is supposed, however, that some four months will elapse before that part of the line is open again to traffic. The present arrangements cause considerable delay to Dorking inhabitants, and passengers generally. For the public convenience, however, the company have altered the trains running between Holmwood and Dorking *[sic]*, eight trains now travelling from Horsham to Holmwood on week days, and seven from Holmwood to Horsham. On Sunday only two trains travel either way.

Dorking & Leatherhead Advertiser
Saturday 4th February 1888
THE BETCHWORTH TUNNEL.

From the official report of the railway authorities we learn that we work of relining and repairing the tunnel will be completed early next month. The work, which is now being carried out under a contract with Mr. J.T. Firbank, will amount to a total cost, including the cost of temporary repairs, to between £18,000 and 20,000, of which sum £10,737 has already been spent. The money for the work is taken from the General Insurance Fund, which has accumulated out of revenue expressly for the purpose of meeting contingencies of this nature.

Surrey Mirror & General County Advertiser
Saturday 11th February 1888
THE BETCHWORTH TUNNEL.

The work of repairing the Betchworth tunnel, between Dorking and Holmwood, is now almost completed. The tunnel collapsed last July, and since that time through traffic on the direct London to Portsmouth line of the London and Brighton Railway system has been suspended. The tunnel, which is nearly half a mile *[sic]* in length, has been entirely recased with brick, and the rebuilding has been carried through in egg shape, so as to secure additional strength. The contractors have almost completed the work, and it has been determined to reopen the line for through traffic on the 1st of March. The cost of rebuilding the tunnel will probably be found not to be far short of £20,000, and in addition to this large outlay, its collapse has, of course, resulted considerable loss to the Company's revenue by the interference caused to the traffic, much of which has been diverted to other lines, although the service of trains has been kept up vîa the Redhill and Horsham route to Portsmouth.

Dorking & Leatherhead Advertiser,
Saturday 18th February 1888
RECONSTRUCTION OF BETCHWORTH TUNNEL COMPLETED.

As the line between London and Horsham, vîa Dorking, will on and from Thursday, March 1st, 1888, be again opened throughout for passenger and goods traffic, the various services of trains between London and Horsham, also between London, Brighton, Littlehampton, Bognor, Chichester, Portsmouth, the Isle of White and intermediate stations, now diverted vîa main line and Three Bridges, will be restored in their entirety, and worked vîa the Dorking line route.

For further particulars passengers must see the time tables, which will be revised for March, April, and May, in accordance with this notice, which will alter and cancel the present publication of the company's train arrangements so far as the month of March is concerned.

To give some indication of the amount of damage that had been done by the collapse, and the work involved to repair it, the tunnel was originally designed and built with slightly curved side walls 1 foot 10½ inches thick, with a segmental arch of a similar thickness.

Left: LB&SCR 2-4-0 tender locomotive No.178, built by Beyer Peacock, Manchester, in March 1864, to a design by John Chester Craven. This engine was hauling the up express from Portsmouth that had emerged from Betchworth tunnel just before its collapse in the summer of 1887.

The detail of the external appearance of this class had been left to the builders, and George Bayer paid particular attention to the shapes of their frames, chimneys, domes and safety valve casings. It should be noted that their various lines are all curves and not radii. Typical of the time, the footplate-men were protected only by a weather-board and not a proper cab. The engine was scrapped in December 1889.

Right: The North portal of Betchworth tunnel, after re-construction. The distinct "egg-shape" of the rebuilt tunnel profile is very evident in the southern portal at the far end. In the foreground, the idiosyncratic approach to track-ballasting used by the LB&SCR gangers is also apparent.

The Standard
Friday 2nd March 1888

Direct through traffic was resumed yesterday from London to Portsmouth on the London, Brighton, and South Coast Railway after having been suspended since July last owing to the collapse of the Betchworth tunnel, which has now been re-cased throughout and approved by the Board of Trade. Some directors and principal officials of the Company went through the tunnel by special train on Monday, and finding everything in satisfactory condition, the ordinary train service commenced yesterday, the first train from Dorking passing through the tunnel at half-past seven.

The Croydon Advertiser & Surrey County Reporter
Saturday 27th August 1888

THE LANDSLIP AT DORKING. – In recognition of the action of Linfield, the ganger who, passing through the Betchworth tunnel just before the falling in of the roof and narrowly escaping being entombed, promptly ran forward and stopped the Portsmouth express, the directors of the London, Brighton, and South Coast Railway have voted the man an honorarium of £10. It is understood that some months will elapse [sic] before the Portsmouth direct line can again be opened for through traffic, as the obstruction is a very serious one, and the landslip will probably involve the necessity of a thorough overhauling of the tunnel.

Yet it is an ill-wind that blows no-one any good. To their credit, the Directors of the LB&SCR eventually resolved to suitably reward the quick-thinking ganger's actions from over a year previously.

In the LB&SCR Engineer's Department staff register for 31st December 1891 there is a James Linfield, aged 54, listed as a ganger at Dorking. If he was the recipient of the reward, it would have made a handsome boost to his wage of 3/9 per day, with one shilling extra for Sunday duty. A further accolade for the Engineer's Department came in the form of the professional recognition given to the LB&SCR resident civil engineer who oversaw the reparation works:

> **Dorking & Leatherhead Advertiser**
> **Saturday 29th June 1889**
> *BETCHWORTH TUNNEL.*
> Mr. George Lopes, BA., Asso. Member Institution, C.E., chief assistant resident engineer of the Brighton Railway Company, has been awarded by the Council of the Institution of the Civil Engineers a Telford premium for his paper on the reparation of the Betchworth tunnel, Dorking.

An abstract of statistics derived from this prize-winning paper appears as Appendix 4. The repairs undertaken by George Lopes and his colleagues have stood the test of time and trains are still able to pass through Betchworth tunnel at 75 miles per hour - a further testament to their engineering prowess, over 120 years later.

However, before leaving this particular subject, Betchworth tunnel has been notable for two further events. Whilst one was decidedly tragic, the other could so easily have been a catastrophe, and both involved a Portsmouth express passenger train.

This is the first, the tragedy:

> **Surrey Mirror & County Post**
> **Tuesday 31st October 1911**
> *SUSPECTED SUICIDE.*
> *DORKING MAN'S DEATH ON THE RAILWAY.*
> What is supposed to be a case of determined suicide occurred at Dorking on Monday afternoon. A local man, named John Ward, living in Church-gardens, was found dead and his body terribly mutilated in Betchworth Park Tunnel shortly after the London to Portsmouth express passed through. The driver appears to have been under the impression that he had knocked down someone, and pulling up at Holmwood, conveyed his suspicions to the station officials. A message was sent to Dorking, the result of which was the discovery of the body. It is supposed that the man jumped in front of the train just before it entered the tunnel, and the body was carried from 50 to 60 feet into the tunnel. An inquest will be held.

And this is the second, a tale of such stupidity and callousness that makes one wonder what the perpetrators could ever have been thinking about:

> **The Portsmouth Evening News**
> **Wednesday 17th May 1922**
> *ATTEMPTED TRAIN WRECK*
> *PORTSMOUTH - LONDON EXPRESS*
> The story of a Portsmouth-London express train's narrow escape from being wrecked by mischievous golf caddies was told at Surrey Quarter Sessions yesterday.
> Stanley James, fourteen, Sidney Windley, fifteen, Alfred Miller, fifteen, and Ernest Lacey, fifteen, all of Dorking, pleaded guilty to endangering the safety of passengers on the London, Brighton and South Coast Railway.
> Mr. Ivan Horniman, prosecuting, said the lads placed on the line at Betchworth Tunnel, near Dorking, two iron bars, each 14ft. long, an iron pipe, an piece of iron metal plate and several large pieces of wood.
> They then sat on a fence to see what would happen.
> The express, travelling at fifty miles an hour, ran into the obstruction. It might easily have been derailed, but it clung to the metals and was pulled up at Dorking.
> The cab of the engine was damaged, the brake-van badly splintered and the engine-driver was hurt.
> The chairman of the Bench, Sir C. Walpole, said that Windley and Miller were evidently the ringleaders. Windley said that if the train went over, he would go down, and if there were any rich people there he would take their money.
> James and Lacey were bound over for twelve months, and Windley and Miller were sent to a reformatory until they were eighteen years old.

LBSCR Class B4 No.52 'Siemens', hauling Portsmouth-bound express, rounds the reverse curves on the approach to the Reigate road bridge and the Betchworth Tunnel. The locomotive was built in December 1899 and named in September 1905. Its umber livery is lined with black bands edged by gilt lines on either side, The engine's original cabside number-plate has been replaced with transfers and the tender is lettered 'LBSC' in gilt block transfers shaded in black, a style that dates from after 1911. Whilst the fireman glances up from adjusting the water flow through the injector, the white feather at the safety valves indicates a full head of steam, ready for the assault on the Holmwood bank. The former spur to the SECR line in the background, used seasonally for hay storage, is visible to the right, behind the train. The duckets [look-outs] at either end of the first vehicle indicate that it is a brake-van.

Chapter 9
EPISODES FROM LATE 19th CENTURY COUNTRY LIFE

The various press cuttings in this chapter require little further comment or explanation, as they really do speak for themselves. They all show how effectively the railway, and in particular the station at Holmwood, have been absorbed into the fabric of the community and its activities. They also show how no lasting damage was caused by the closure of the line following the collapse of Betchworth tunnel and quite how responsive the LB&SCR was to the demands of its passengers.

Surrey Mirror & General County Advertiser
Saturday 7th June 1884
MARRIAGE OF MISS PENNINGTON AND MR. FRANCIS

The pleasant little villages of Coldharbour and Ockley were yesterday the scene of what has for some time past been a source of pleasurable anticipation to the inhabitants, viz., the marriage of Miss Evelyn Pennington, eldest daughter of Mr. Frederick Pennington, MP of Broome Hall, Holmwood, Dorking, to Mr. Norman Arbuthnot Francis of the 18th Royal Irish Regiment. The ceremony was performed at Ockley Church, which had been beautifully decorated by the skilful hands of Mrs. and the Misses Calvert, assisted by the Misses Lee Steere … …

After the ceremony the happy pair and guests returned to Broome Hall, amidst a shower of flowers and rice, where a recherché breakfast was partaken of in a marquee …

At half-past three o'clock the newly wedded couple left the Holmwood station en route for Scotland, to spend their honeymoon. … …

The Daily News, Wednesday
26th September 1888

The National Sunday League announces that their last excursion will be next Sunday to Leith Hill (Holmwood Station) and Horsham. Trains leave London Bridge and Victoria at 9.40.

Dorking Advertiser, Saturday
24th August 1889
THE I C U MINSTRELS' EXCURSION.

On Monday morning when the specially chartered, well arranged train drew up alongside the platform of the Dorking station on the Brighton line, "spick and span" for her journey down South, it was seen that the extensive arrangements the I C U Minstrels had made for the reception and conveyance of their patrons to that empress of watering places, Eastbourne, were in no way superfluous, or on a scale too elaborate for the event. The airy and comfortable electric-lighted carriages were all requisitioned, and when every one of the excursionists had been picked up it was found that not a square inch of room was wasted. Arrangements were made by the Minstrels with the obliging station-master, Mr. Morrison, for the conveyance of between five and six hundred persons, and it was found, on Monday morning when the train drew up at the platform that there just about this number awaiting its arrival. For the minstrels and their friends a comfortable saloon carriage was set apart, whilst the rest of the train consisted of the most airy, and comfortable carriages on this enterprising company's line. Shortly after eight o'clock the train steamed off, and after picking up a few additional passengers at Holmwood and Ockley stations, an uninterrupted journey to Eastbourne was enjoyed. … …

Surrey Mirror & General County Advertiser
Saturday 7th June 1890

FIRE. – On Thursday evening a very large haystack belonging to the London, Brighton, and South Coast railway Company, and standing on their property near Betchworth tunnel, caught fire, presumably from the sparks of a passing engine. The stack burnt freely and was still alight yesterday (Friday) morning.

Dorking Advertiser, Saturday 13th July 1889
A DAY AT THE SEASIDE.

The members of the Dorking Church of England Bands of Hope, together with the members of the Abinger Band of Hope, had their annual excursion to the seaside on Monday. Littlehampton was the selected resort, and here, in spite of the drizzling rain which fell with annoying persistence during a greater part of the day the large number of children and adults who journeyed thither under the auspices of the excursion, passed eleven hours away in tolerable enjoyment. ... ...

The combined party numbered nearly 900. This was many in excess of any number anticipated, and the obliging and willing station master (Mr. J. R. Morrison) experienced some difficulty in finding sufficient accommodation for the travellers. However he succeeded in safely seating them in a train of 19 coaches, and amid the huzzahs of those who were left behind the heavily weighted train eventually steamed off. A few more persons were picked up at Holmwood, after which the journey to Littlehampton was quickly made. ... ...

Left: A postcard of Broome Hall, franked on 8th October 1908. The village name 'Ockley' has been blotted out in ink and 'Holmwood' substituted.
Right: A postcard, almost certainly of similar vintage, showing the porch of Ockley Church.

Dorking Advertiser, Thursday 2nd August 1894
A BRIGHT DAY AT NEWDIGATE
HOUSE WARMING, AND SILVER WEDDING FESTIVITIES

Mr G.O.M. Herron a few years ago purchased Newdigate Place Farm. He had a large mansion built of red bricks covered with red tiles. Above the line of the first floor the mansion is weather tiled in the same colour. The timber used in the building was well seasoned oak, grown on the estate. The carriage drives and garden walks were made with "dencher" clay burnt for the purpose, in a field close by. At present they are in a very rough state. Until last week Mr. and Mrs. Herron resided at "Ockley Lodge", another nearly new house on the estate, which was built by the late Mr. William Farnell Watson. To celebrate their taking possession of the new mansion, and also to commemorate their "Silver Wedding day" - July 21st last - Mr. and Mrs. Herron invited 150 friends and neighbours to a round of festivities at their new residence on Saturday afternoon last.

A special train brought many friends from London to the Holmwood Station, whence they were conveyed in twelve carriages to Newdigate Place. They returned to Holmwood Station shortly after midnight. The school children from Newdigate and Charlwood parishes were invited; those from the latter place were brought in waggons, decorated with flags and green boughs. All the girls had been provided by Mr. and Mrs, Herron with bright coloured print frocks and white straw hats. Leaving the wagons a short distance away, they marched to the lawn signing "Auld Lang Syne". They then danced and sang, and – so our correspondent writes – "performed many interesting antics" around a Maypole on the lawn. They then went to Boothland Farm, near by, and had tea.

An Orchestral band, provided by Mr. Groves of Tunbridge Wells, discoursed sweet music under a large Chinese Umbrella Tent at the side of the lawn. While the guests had dinner, the Chinese lanterns and fairy lamps which had been fixed to the trees and shrubs around the lawn were lighted. Presently the children returned, singing and bearing aloft lighted Chinese lanterns. They distributed themselves on the outside of the lawn, singing and waving their lights. Viewed and heard from the Terrace, the sight and sounds suggested fairyland. Upon leaving, the youngsters made the welkin ring with many and loud "Hurrah's". All rode home on waggons. While this was going on, the bandsmen had their supper, and then the large entrance hall was cleared dancing, which was kept up till after eleven. The house is fitted with electric bells and electric light; only a few lamps had been fixed in the Hall. These were alight - sometimes - but the light was quite unsteady, and once or twice quite went out, causing great fun. Talking of lights - a pretty spectacle of the night hours was the burning of coloured fires on a pond near the house. ... ...

The horses who found so much work to do that day were accommodated at the Home Farm, and at Ockley Lodge, the coachmen having supper provided for them at the latter place. Mr. and Mrs. Herron's servants and helpers were to have their treat on Monday evening last, so none were forgotten.

A short distance north east of the house is a water tower, provided with an engine to pump water and generate electricity. The engine was only fixed on Saturday morning.

Dorking Advertiser
Thursday 5th July 1894

EXCURSIONISTS AT THE HOLMWOOD - Last Thursday about 250 persons from the Rev. C.H. Gould's former parish at Portsmouth, came by an excursion train to The Holmwood Station, on a visit by invitation to Mr. Gould, now vicar of The Holmwood, who entertained them to tea. A special service was held in the Church. The visitors rambled about the Vicarage grounds and elsewhere when not wandering over the ever delightful common.

Dorking Advertiser, Saturday 14th May 1892
DORKING POLICE COURT

DRIVING. William Jennings, Farmer, of Hatchetts Farm, Newdigate, was charged with furious driving on the highway at Holmwood station on May 3rd. – P.C. Bourne gave evidence. Defendant said he pulled up when the constable called out to him on the road. His mare was one of fresh courage, and would break into a canter sometimes going up hill. It was a little nervous when near a railway station. – Fined 7s. 6d. and 12s. 6. costs.

**Dorking Advertiser
Saturday 29th November 1890**
ABOUT LOCAL PEOPLE

As we briefly stated in our last issue the funeral of the late Mr. W. C. Cazalet took place on Thursday, the 20th inst., at South Holmwood. The funeral car, drawn by four horses, and followed by three mourning coaches and numerous carriages containing many of the mourners left Grenehurst about two o'clock. Before entering Capel the cortege was joined by many local tradesmen. The greater number of those living on the estate, and in Ockley, who attended the funeral travelled by train to Holmwood station. Whilst the body was borne through Capel the bells of the village church rang muffled peals, as they did again in the evening. At Holmwood station several relatives and friends entered carriages whilst the procession halted to enable them to join its ranks. It then continued slowly on its way to the churchyard. The coffin was borne by eight labourers from the estate. It was covered with wreaths. The vicar, the Rev. E. D. Wickham, conducted unaided the whole service. The first portion was conducted in the sacred edifice. The body of the church was filled by the mourners and others fain to pay the last token of respect to one so worthy. The coffin was then borne into which it was lowered, and finally placed in the vault in which lie the remains of Mr. Cazalet's mother and two of his sisters. Fine weather favoured the procession during its progress and the return of those who had come on foot. ... ...

**The Dorking & Leatherhead Advertiser
Thursday 5th December 1895**
*HOLMWOOD
WAS HE MAD OR ONLY SAVAGE.*

Great excitement was caused in the neighbourhood of Holmwood Station last Monday afternoon, when it became known that a dog belonging to Mr. Charles Lipscombe, coal merchant, had bitten its owner in the shoulder, and also bitten the hand of a young man in his employ, named Ansell. Mr. Lipscombe's wound was slight; Ansell's more severe. The dog was shot – executed without trial. Their injuries were attended to by Mr. Lee Jardine, surgeon, of Capel.

**The Dorking & Leatherhead Advertiser,
Thursday 13th June 1895**
10½ HOURS AT THE SEASIDE.
WEDNESDAY, JUNE 19th
LEATHERHEAD CONGREGATIONAL
WESLEYAN SUNDAY SCHOOL
EXCURSION

———————

**Leaving Ashtead 7.30;
Leatherhead 7.35;
Dorking and Holmwood (at usual intervals), Ockley at 8 o'clock**

Tickets obtainable from
Messrs. Batten, Bulpin, Dearle, Kirkland, Mould, Hewett, Shew and Jenden at Leatherhead;
Doubleday and Nicklin at Dorking; Chitty at Bookham;
West at Effingham;
Weeden and Palmer at Capel
Deane at Beare Green.

Also at each station.
3s. ; under 15, 1s.6d.

One of the attractions of a property advertised in 1895 by the letting agents Messrs. Vigers [see overleaf], was its proximity to Holmwood station. Another was having six packs of hounds within reach. These provide useful hints about the likely wants of a prospective tenant with business interests in the metropolis, but wishing to take up the life of a country gentleman. In the next Chapter, one of the more unexpected services provided for its wealthy patrons by the LB&SCR is explored in some detail. This involved the transport of horses and hounds on a hunting day, both for the Surrey Union Hunt itself and for any private individuals who wished to get to the meet, together with their hunters.

The Morning Post, Monday 29th April 1895

HOLMWOOD (near Dorking). – BEAREHURST.

Good FAMILY RESIDENCE, situate in a charming position, five miles from Dorking, one mile from Holmwood Station (L.B. and S.C. Railway), one hour from London by express; the house contains large hall, three reception rooms, billiard-room, 14 bedrooms, bathroom, servants' hall, and offices; good cellars, two staircases; heated throughout by hot water; two acres of garden and grounds; two good cottages; stabling for seven horses, two rooms for grooms; 20 acres of pasture if required; six packs of hounds within reach; to be LET on Lease.

Apply to
Messrs. VIGERS and CO.,
4 Frederick's-place,
Old Jewry,
London, E.C.

Designed by William Stroudley, D1 class 0-4-2 tank engines were ubiquitous on the line through Holmwood from 1873 until after the Grouping in 1923. As no clear photograph of this class at the Holmwood station has been found, shown [left] is No.285 *'Holmwood'*, probably at Epsom shed, whilst [right] No. 297, previously named *'Bonchurch'*, arrives in Ockley with a train from Holmwood in 1922.

Locomotives No. 264 *'Langston'* [built by Neilson & Co, Glasgow 1882] and No.285 *'Holmwood'* [built at Brighton in 1879] again probably on Epsom engine shed, pre-1894. This black & white image does not do full justice to their 'Stroudley Improved Engine Green' livery [actually a golden ochre colour], offset by areas of dark olive green and crimson lake to further enhance its lustre. These accent colours were finely lined out in red and white, with a thicker line of black. The frames are crimson lake, lined with yellow and vermillion. Cab roofs were painted white and the cast brass number-plates had a dark blue infill. Both locomotives were withdrawn in November 1926.

Chapter 10
HUNTING & THE RAILWAY

Following the opening of the line in 1867, it did not take long for the Surrey Union Hunt to realise the opportunities that the railway offered for the transport of horses and hounds from its Kennels in Fetcham. Indeed, the approach to the station goods yard made an excellent venue for a meet:

> **The Morning Post, Saturday 19th December 1874**
> *HUNTING APOINTMENTS*
> Surrey Union - Monday, Effingham Village; Wednesday, Pickhurst Chiddingfold; Thursday, Perbright [*sic*]; Saturday, Holmwood Station, 11.

A special train would often be run for this specific purpose on a hunting morning from Leatherhead, the closest station to the Hunt Kennels, to the most convenient station for the meet. The meet card [below left] is typical of the period and shows Holmwood as a destination on 31st December 1898. This frequent use of special trains by the Surrey Union was sufficient to induce the editor of "The Tatler" to produce a supplement illustrating such an event on 21st March 1906 [below right]. Enlarged images from that piece are shown opposite and the accompanying text appears on page 46.

SURREY UNION HOUNDS

WILL MEET AT 11 o'Clock.

Tuesday,	Dec. 13th	Claygate.
Thursday,	„ 15th	Effingham.
Saturday,	„ 17th	Cranleigh Lane End, **11-15**.
		(Special Train, Leatherhead, 9.30.)
Tuesday,	„ 20th	Bagden Farm.
Thursday,	„ 22nd	Forest Green.
		(Special Train, Leatherhead 9·30.)
Saturday,	„ 24th	The Swan, Leatherhead.
Monday,	„ 26th	Mickleham Village.
Thursday,	„ 29th	East Clandon.
Saturday,	„ 31st	Holmwood Station.
		(Special Train, Leatherhead, 9-30.)

Special Train does not stop anywhere to pick up Boxes.

Arthur Labouchere, *Cobham*.

Hunting Reynard By Rail and Horse

Above: The mounted field begins to gather at the end of the station approach road, in front of the white palings that mark the entrance to the goods yard. The group of figures standing by the locomotive, whilst the top doors and ramp of the first vehicle remain shut, suggest that hounds have not yet been unloaded - this would be the last task in the sequence of events. There are at least nine vehicles in the train.

Left: Hounds spill out of their van. It looks as though all the station staff have come to assist the whipper-in. Is the locomotive a Stroudley class D1 0-4-2 tank engine?

Right: Following their careful unloading from the train, all the Hunt horses are being led towards the side gate of the down platform. To the right, an alert groom keeps an eye on things.

The final photograph shown in the Supplement [above] shows the Hunt leaving the meet at, rather aptly, 'The Fox' public house at Norwood Hill, a hack of about 4½ miles from Holmwood station. Taking into account the likely time taken up by the various elements of the journey and the expeditious unloading of the train, the entourage probably left the station yard probably no later than 10-00am.

Nevertheless, after the Great War and before the electrification scheme in the summer of 1938 added a 650 volt live rail and closed most farm accommodation crossings, the Surrey Union Hunt did seem to have something of a 'devil may care' attitude towards railway safety [see page opposite].

The Tatler - Supplement No.247
Wednesday 21st March 1906
Hunting Reynard by Rail and Horse
Reynard is nowadays not infrequently hunted by rail as well as horse. The photographs above give an excellent idea of the manner in which the railway may help the sportsman in covering long distances. During a run of the Surrey Union Foxhounds the whole company travelled from Leatherhead to Holmwood below Dorking: here the hunts men, women, horses, and hounds were ready to start after a space of ten minutes.

Thomas' Hunting Diary 1905-1906
pages 106-107
Railway Arrangements for Hunting Men
London Brighton and South Coast
No reduction is made in the ordinary fare for gentlemen travelling for hunting purposes – but for the convenience of gentlemen hunting in the vicinity of the line, return tickets are issued for horses when accompanied by huntsmen or groom, at the reduced charge of a fare and a half, available on day of issue only and at owner's risk.

**Horse & Hound Saturday
5th February 1927**

... and on we went to Green's farm, through The Views to Henfold and Brexells [sic], and over the railway just north of Holmwood Station (where a good engine-driver pulled up his train) to Holmwood Park. ...

The Surrey Union Hunt Accounts for 1936-37
Paid to Engine Drivers - stopping trains £ - 10s 0d

The Surrey Union Hunt Accounts for 1937-38
Paid to Engine Drivers - stopping trains £2 10s 0d

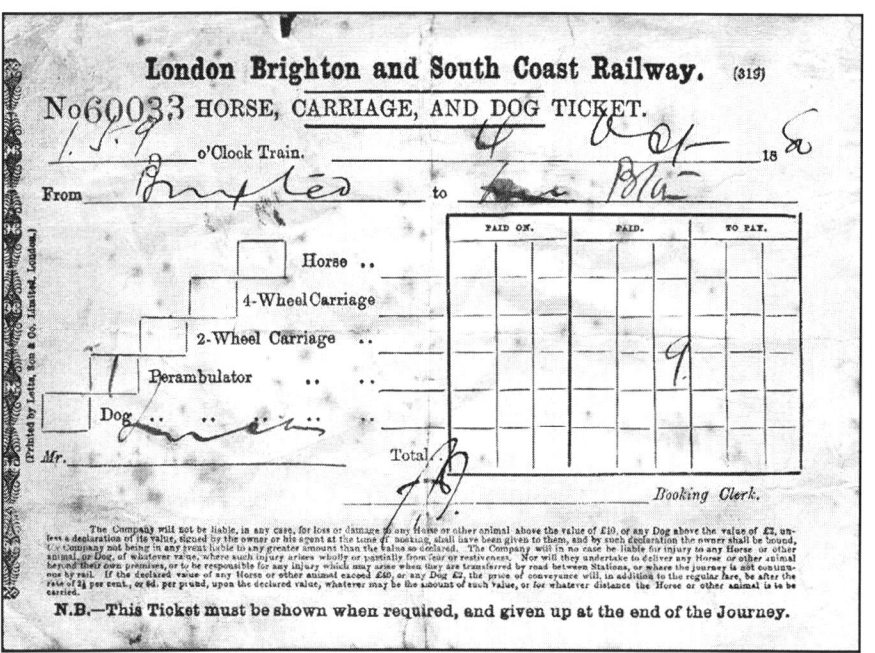

The Surrey Union Hunt crosses the railway on a farm accommodation crossing, just to the south of Holmwood station in December 1936. The Holmwood up distant signal appears to be off, but no-one in the mounted field seems to be unduly concerned.

A Horse, Carriage and Dog ticket issued by the LB&SCR on 4th October 1880. Alas, rather than for a hunter or a pack of foxhounds, it is made out for a perambulator!

Chapter 11

HOLMWOOD JUNCTION?
THE HOLMWOOD & CRANLEIGH RAILWAY

In the London Gazette of 25th November 1898, a notice was published announcing the introduction of the London Brighton and South Coast Railway (Various Powers) Bill into the parliamentary session for 1899. Amongst various other things, this proposed various changes at Holmwood and a completely new railway line from Holmwood to Cranleigh:

> A railway commencing in the parish of Capel by a junction with the Company's Horsham, Dorking, and Leatherhead Line, at a point thereon 19 chains, or thereabouts, measured in a south-westerly direction along the said railway from the booking office at Holmwood Station thereon, and terminating in the parish of Cranleigh by a junction with the Horsham and Guildford Line of the Company at a point thereon 3 chains, or thereabouts, measured in a south-easterly direction along the said railway from the booking office at Cranleigh Station thereon, which said intended railway will be wholly situate in the county of Surrey, and will pass through or into the following parishes and places, or some of them, viz.: - Capel, Ockley, Ewhurst, Hambledon, Wotton, Abinger, and Cranleigh.

The new railway line that the LB&SCR intended to build, linking its stations at Holmwood and Cranleigh, would have been 8 miles and 31 chains in length. This would have reduced the distance by rail between Cranleigh and London from 50 miles to 39 miles 5 chains. The cost of this new line would have been £132,436, to which a further £13,787 would need to be added for the widening of the existing lines and improvements to the station at Holmwood. It seems to have been the intention to build an intermediate station at Ewhurst, with the possibility of another at Forest Green.

The desire for this new railway line came from the residents of Cranleigh, who were dissatisfied with their passenger service to London, via Guildford, and which often involved an inconvenient connection at Guildford with trains to Waterloo run by the London and South-Western Railway. This clamour was augmented by farmers, brick-makers and other commercial interests, including the Cranleigh Land Company. The inhabitants of Ewhurst and Forest Green also passed resolutions in support of the proposed new line.

At the parliamentary hearings, the arguments for the line were probably best summarised in the evidence given by the Venerable Archdeacon Sapte, the Rector of Cranleigh for the previous 53 years, who had not only suggested the new line to the LB&SCR in the first place, but also had no doubt that fresh industries would spring up in the neighbourhood, property values would improve and the numbers attending Cranleigh School, of which he was a trustee, would increase. A further insight into how the new railway could improve the commercial prospects of the district came from Mr Frederick Elliot, the proprietor of a firm of timber merchants in Cranleigh, who told the House of Commons Committee that he could never send goods to London via Guildford as the rate by that route was 9 shillings per ton, whereas via Horsham it was only 5s.2d. Furthermore, he had been promised that the rate would be slightly lower still by using the new line.

Ranged against this widespread support at the Cranleigh end of the new line there was some pretty stiff opposition to the proposal from four landowners in the Holmwood area: Colonel Charles Mortimer of

'Wigmore', Colonel Calvert of Ockley Court, Mr Augustus Perkins of 'Oakdene' and Mr Alexander Hargreaves Brown MP of Broome Hall. Their objections were based on the visual intrusion of the new line into the rural views from their respective mansions, the despoliation of their agricultural and sporting estates and the threat posed by the uncontrolled roaming of day-trippers. In the view Mr Hargreaves Brown, if public necessity could be claimed he would willingly yield to it, but the proposed line was unnecessary and really of little value.

The editorial piece [on the right] from the Dorking Advertiser & County Post, sycophantic and condescending as ever, summarises the debate and records its outcome.

Over a hundred years later, it is perhaps a futile exercise to try to speculate what might have become of the villages of Ewhurst and Forest Green, or its effect on Beare Green, had this proposed railway line been built. However, just as the line from Christ's Hospital through Cranleigh to Peasemarsh Junction failed to survive the Beeching cuts of the 1960's, it is unlikely that this line would have fared any better – even assuming it would have lasted even that long. Yet what an asset the track bed of this line would have proved in the current age of the Sustrans cycle routes and long distance footpaths, just as other disused railway lines have become in rural Surrey.

Above: Alexander Hargreaves Brown, Esq M.P.

The Dorking Advertiser & County Post
Saturday 6th May 1899
EDITORIAL
Holmwood to Cranleigh

It is, perhaps, only natural that those who are most deeply interested in the maintenance of the rural beauty of the district which lies between The Holmwood and Cranleigh, among whom are, of course, to be included the majority of the wealthy and cultured classes, should most strongly oppose the Bill which had for its object the connection of the two places and the opening up of the intervening district by the line of railway which the Brighton Company have been seeking powers to build. No one possessing a rural estate of great beauty could reasonably be expected to look with equanimity upon the introduction of the tripper element, and in consequence the opposition which the Bill met with was exceedingly strong. It included practically the whole of the residential population, while it is hardly too much to say that the only support which the Bill received was from the trading community and others interested in the development of Cranleigh. Dorking had, indeed, passed an academic resolution in its favour, but even Dorking firmly declined to welcome the line within its borders. It is not therefore, surprising that the Select Committee of the House of Commons, after lengthy enquiry, which will be found reported in our columns, should have found yesterday that the preamble of the Bill could not be regarded as proved so far as this portion is concerned. For the present, therefore – but in all probability for the present – the project has come to an end, but early railway development is inevitable. The district which the proposed line would have traversed is but sparsely populated, but the time for opening it up by railway communication will, of course, come sooner or later. The railway company have done everything in their power to serve the interests of the district, and in all probability their action was due mainly to a laudable desire to meet the wishes of Cranleigh. So far they have discharged their public duty.

Chapter 12

THE DEATH OF QUEEN VICTORIA

Queen Victoria died at Osborne House, on the Isle of Wight, on Tuesday, 22nd January 1901. However the Queen's ailing condition was made known within the extended royal family beforehand and many of them were beside her at the end. Particularly notable amongst those present, at least as far as this narrative is concerned, were the Queen's sons, the Prince of Wales and the Duke of York, and her eldest grandson, Emperor Wilhelm II, the German Kaiser.

It is thought that the late Queen was not particularly enamoured with the LB&SCR, despite the company possessing the nearest railway terminus to Buckingham Palace and, from Victoria station, a main line leading directly to Portsmouth harbour and its maritime services to the Isle of Wight. In marked contrast, the new King, his advisers and entourage were not slow to take advantage of this convenient route:

> **The Dorking Advertiser & County Post**
> **Saturday 26th January 1901**
> *DORKING*
> ... It was the privilege of a few who were on the Dorking platform of the London and Brighton line on Monday to catch a glimpse of His Royal Highness, who, as the Prince of Wales passed through to Osborne, returning to London on Wednesday morning by the same route, as the King of England. ...

The same would obviously be true in the case of Holmwood station, but on that same Wednesday a casual observer on its up platform would have been a yet more privileged witness to this most unlikely and astonishing spectacle:

> **The Dorking Advertiser & County Post**
> **Saturday 26th January 1901**
> *HOLMWOOD*
> ... It is a matter of interest to this village that the Royal train stopped here on its way to London on Wednesday, and the King, accompanied by the German Emperor and the Duke of York alighted on the platform for the purpose of despatching a telegram. ...

Not surprisingly, this same route was traversed by the funeral train that brought the late Queen's body back to London from Gosport, where it had laid overnight after being brought across the Solent from Osborne. Many people turned out to pay their last respects as the train passed by:

> **The Dorking & Leatherhead Advertiser**
> **Saturday 9th February 1901**
> *CAPEL*
> ... A large number of persons walked over to the side of the railway to see the special train pass conveying the remains of the late Queen from Portsmouth [sic] to London. ...

The arrangements for working Queen Victoria's funeral train on Saturday, 2nd February 1901 were set out in a four-page Special Traffic Notice, published the day before. It was issued jointly by the London and South Western Railway and the London Brighton and South Coast Railway, through their respective General Managers (Charles J. Owens and William Forbes) and Superintendents of the Line (Sam

Fay and D. Greenwood). It contained details of timings; the vehicles in the train; head codes to be used on the locomotives; precautions to be observed, together with specific instructions to station masters, guards, signalmen, platelayers and level-crossing keepers.

The Royal Train was to be comprised of eight vehicles and, on leaving Fareham, they should run in the following order:

[1] Brake Van, [2] Saloon, [3] Funeral Car, [4] Royal Saloon, [5] Saloon, [6] Bogie First, [7] Bogie First, [8] Brake Van.

The Funeral Car was a Great Western Railway clerestory-roofed saloon, carriage No. 229, from that Company's Royal Train, stripped out and prepared appropriately. Vehicles 1, 2, 4, 5 and 8 were the usual LBSCR Royal Train, whilst vehicles 6 and 7 were the two coaches that sometimes augmented this rake should the occasion arise.

Running the train over the lines of two rival railway companies made the whole operation rather more complex than it need have been. The chosen route from Clarence Yard, Gosport to Victoria Station in London involved a locomotive change and reversal of the train direction at Fareham. Thus the front carriages on leaving Gosport would be the rear carriages on arrival in London. Initially this subtlety was overlooked and it took some time and, one suspects, a great deal of tact to persuade the royal and very important personages to change their seats from one end of the train to the other prior to departure. This caused a delay of eight minutes. A further complication arose at Fareham when the L&SWR locomotive, A12 class 0-4-2 No. 555, was taken off one end of the train and the LB&SCR locomotive, class B4 4-4-0 No.54 'Empress', attached at the other, as the driver of the latter was unable to make a brake. Eventually, after the intervention of a fitter to make the necessary repairs, the Royal train left Fareham 10 minutes late. It was well known that the new King disliked unpunctuality, so the footplate crew were instructed to make up the lost time. Inevitably, this produced some lively running and a sparkling exhibition of locomotive handling by driver Walter Cooper and fireman F. Wray.

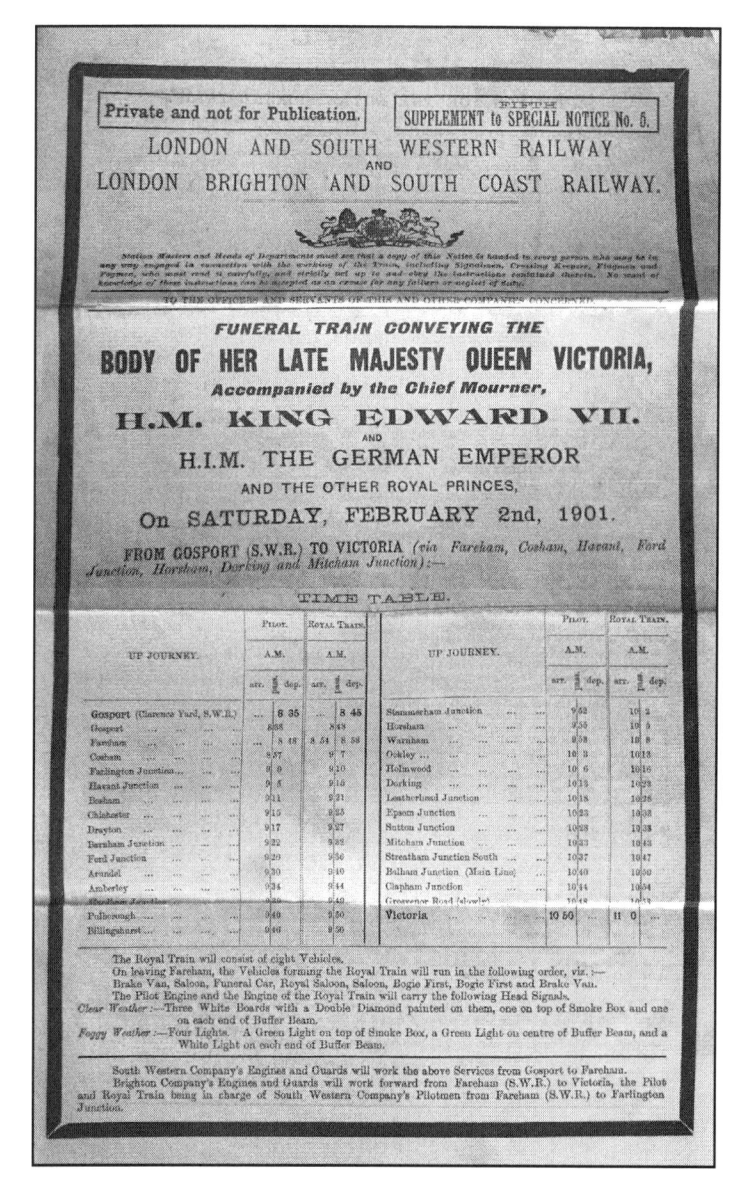

The Special Traffic Notice for Queen Victoria's funeral train, issued jointly by the London and South Western Railway and the London Brighton and South Coast Railway.

No. 13.—Funeral Procession of Queen Victoria.
Copyrighted 1901 by M. E. Wright.

A speed of 80mph is thought to have been achieved between Havant and Ford Junction, but the most interesting, and subsequently controversial, part of the journey was the descent of Holmwood bank. Here, some railway officials travelling in the train estimated a maximum speed of 92 mph was reached and the rough riding of the GWR saloon caused some alarm, as it was though that the Queen's coffin was in danger of falling off the catafalque. The disquiet was increased by the knowledge that the reverse curves leading into Dorking at the foot of the bank carried a 30 mph speed limit. Subsequent analysis of all the evidence, including the fact that the both the LBSCR Locomotive Superintendant, Mr Robert Billinton, and his Outdoor Locomotive Superintendant, Mr J. Richardson, were on the footplate of No.54, it seems unlikely that the speed exceeded 75 mph and an effective use of the Westinghouse air brake in Betchworth tunnel would have reduced speed to a safe level on the approach to Dorking station. In the end, London was reached 2 minutes early - a point not lost on the Kaiser, who thought it a remarkable achievement by so small a locomotive, especially when compared with those used on German railways.

Taking all of this into account, it is not entirely clear when the funeral train appeared at Holmwood. Doubtless the pilot engine, 'Sirdar', ran pretty close to time, but it seems likely that those standing in the bitter cold were kept waiting past the scheduled time of 10.16 am before the Royal train passed through at speed.

This press cutting provides a very accurate account of the events of that morning and seems to have been written with a clear insight into the contents of the Special Traffic Notice and the railway operation generally:

Opposite Left: His Imperial Highness, Emperor Wilhelm II, the German Kaiser.

Opposite Right: The Royal mourners. This is an image taken from a commercially produced series of stereoscopic slides showing Queen Victoria's funeral cortège as it passed through Cowes. It was on its way from Osborne to the Royal Yacht "Alberta", for the passage from the Isle of Wight to Royal Clarence Yard, Gosport on the mainland.

The Surrey Mirror, Saturday 8th February 1901
THE JOURNEY FROM GOSPORT TO LONDON

... ... and it was not till Saturday morning that it was definitely made clear that the London and Brighton system would be adhered to all through [sic], and that the journey from Gosport to Victoria would be made vîa Horsham, Dorking and Epsom. The news spread rapidly, with the result that all the railway embankments and bridges, and, in fact, all points of vantage from which a close view of the Royal train could be obtained, were before ten o'clock studded by hundreds of people, in spite of the rain that was falling. Thanks to the courtesy of Mr Holdaway, the Company's obliging representative at Dorking, many of the public, including a number of prominent residents, were permitted to enter the station precincts, and to take up positions on both the up and the down platforms. Elaborate arrangements were made to guard against any untoward incident. Supt. Alexander was present with a strong body of the Surrey Constabulary, including an additional force from Guildford, and it was noticed that each of the men wore a mourning band around his coat sleeve. Constables were also on duty guarding the bridges and tunnels, while platelayers were placed all along the line to signal the approach of the train by flags. These arrangements were in force over the whole of the route between Gosport and Victoria. The Royal train, which was timed to leave Gosport at 9.30, was due at Dorking at 10.23. The pilot engine passed through at exactly 10.15, followed a few minutes after the scheduled time by the Royal train. At its approach heads were reverently bared, and though the assembled watchers gained but a passing glimpse of the swiftly-moving train, the deep feeling of reverence shown all along the route was more eloquent than words can tell of the deep love and attachment for her who was passing to her final resting place. As the train approached Dorking it was necessary, in order to safely negotiate the sharp curve to slow down to about 30 miles an hour, at which speed the train passed through the station. The train consisted of eight vehicles, in addition to the engine - two brake vans, two ordinary saloons, the funeral car, a Royal saloon and two first-class bogie carriages. The hand-rail of the engine was draped with purple and white, while on the top of the smoke-box was a crown, and the three discs had double diamonds in black painted on them. The interior of each saloon was upholstered in white corded silk with a frieze combining national emblems. In the funeral car the tables had been removed, and in their place stood the bier. This was about four feet high, and was draped in purple, with white rosettes. From the brass curtain rods hung festoons of purple, with white rosettes to correspond with those on the bier. In each corner of the car was a chair covered in purple cloth, on which sat State officials in charge of the coffin.

The same feeling of deep reverence and devotion was manifest throughout the Leatherhead district, where the residents gathered in hundreds all along the line to pay their last respects to the mortal remains of her whom they all so loved. Thorncroft Bridge, and the vicinity of the cricket ground seemed especially favoured spots from which to view the train, while many chose the Station Road and the Kingston Bridge.

The public were rigidly excluded from the railway station, the application for permission for members of the Urban Council to assemble on the platform meeting with a polite refusal.

Top: Queen Victoria's funeral train near Carshalton, en route to London. The late Queen's coffin was being carried in the third vehicle. This scene would have been replicated in the fields adjoining Holmwood station. Note the gentlemen removing their hats as the train passes by.

Bottom Left: The locomotive that drew the funeral train on the LB&SCR, class B4 4-4-0 No.54 'Empress'. The engine first entered service in February 1900. Note the subdued funeral drapes around the boiler and smoke-box hand-rails, whilst at the foot of the chimney, a regal crown is carried on the top lamp-bracket.

Bottom Right: Sister locomotive, No. 53 'Sirdar', ran as pilot 10 minutes before the funeral train. As the engine appears to be in 'photographic grey' livery, it is probable this photograph was taken soon after its construction in January 1900.

The difference in the behaviours of the two Station Masters reported in the previous extract should be noted. In his defence, the response from the Station Master at Leatherhead is not quite as churlish as it may seem. On the other hand, that of Mr Holdaway in Dorking is really rather inexplicable bearing in mind the specific instructions that had been issued to all railway employees.

The Special Traffic Notice is quite explicit:

STATIONS TO BE KEPT CLEAR AND PRIVATE

32. All the stations must be kept perfectly clear and private during the passage of the Royal Train, and no persons (excepting those properly authorised, Passengers travelling in the opposite direction, the Company's Servants on duty, and the Police at those stations where their services are required) are to be admitted to any of the stations on the route. The Servants of the Company are to perform the necessary work on the Platforms without noise, and no cheering or other demonstration must be allowed, the object being that the Royal Party shall be perfectly undisturbed during the journey, all the stations being kept perfectly clear and private.

So, was the Station Master at Dorking admonished by his superiors in the LBSCR? But perhaps more to the point, did the Company's Servants at Holmwood behave as instructed? Let us hope so.

Left: A further portrait of No.54 'Empress' specially prepared for the Royal funeral train.

Above: A commercially produced glass 'magic lantern' slide showing an artist's impression of the interior of the saloon that carried Queen Victoria's coffin from Portsmouth, after its arrival in London.

55

Chapter 13

EPISODES FROM EARLY 20th CENTURY COUNTRY LIFE

The new Century started much as the old one had ended, with Holmwood station at the very heart of community activity seeing a variety of jolly jaunts, returning Boer War heroes, a sorry litany of accidents and fatalities, miscreants and bank holiday trippers:

> **The Dorking Advertiser & County Post**
> **Saturday 28th July 1900**
> *A DAY IN THE COUNTRY*
>
> It is most pleasing to record that another of Southwark's princely merchants, Mr. A.F. Perkins, of Oakdene, Holmwood, has enabled about seventy-five of the Almspeople of St. Saviour's, now living in the almshouses at Norwood and Dulwich, who have been reduced in position by circumstances over which they had not any control, to spend a most pleasant day in the country, at the seat of Mr. Perkins, on Wednesday of last week. The party, ... , assembled at the Almshouses, Hamilton-road, West Norwood at 8.30 a.m., and were conveyed to Streatham by waggonettes, and thence by train to Holmwood Station, where they were met by Mr. Perkins. At Oakdene, the most bountiful provisions were made. After the preliminary luncheon a number of the visitors were taken for drives affording the opportunity of admiring the beautiful scenery surrounding Leith Hill and Ockley. At 1.30 the party sat down to luncheon, presided over by Mrs. and Miss Perkins, and Miss Clarke, whose indefatigable labours and kindly sympathy spontaneously won the hearts of the visitors. During the afternoon more drives were taken, and the Dorking Town Band arrived upon the lawn and rendered some thoroughly enjoyable music. Sports were indulged in; ... Messrs. Hill and Newton led off with an egg and spoon race, but the chief interest was centred in the races between the old people, both sexes vying with each other to gain the first and second prizes, which the hostess had ready to bestow. The step dances for men and women were particularly attractive. Miss Perkins completed the enjoyment of the afternoon by giving an exhibition of the art of measured steps. That lady also photographed the Wardens, and seven of the oldest almspeople, whose aggregate age was 600 years, and there were many others of about 80 years deemed too young deemed too young to be included in the portraits of the ancients. Tea being over, Mr. Perkins responded to a vote of thanks moved by Mr. Lawrence (an inmate) and explained the great pleasure which Mrs. Perkins and himself took in doing what they could to brighten the condition of the deserving and expressed the hope that the almspeople would live long enough to enjoy another meeting of the same character. Mrs. Perkins then presented each visitor with parcels of tea and tobacco, and the band played the National Anthem, bringing the most delightful day to a conclusion. The almspeople arrived home safely about 8.30 p.m.

> **The Dorking Advertiser & County Post**
> **Saturday 12th January 1901**
>
> ACCIDENT AT THE STATION. – A young man named Wheeler, a porter at the Holmwood Station, met with a serious accident on Tuesday night. He lives in Leslie-road, Dorking, and at the conclusion of his day's work endeavoured to enter the Dorking train when in motion. He missed his footing, was caught by the coupes of the guard's brake, and thrown down between the platform and the carriages. Luckily, he fell between the brake van and the next carriage on to the line, thus avoiding severe crushing between the footboard and the platform. The accident was fortunately observed by the guard, who instantly applied the Westinghouse brake, with the result that the train was at once pulled up. Wheeler was then extricated, and it was found that although the wheels had not passed over him he had received a terrible injury to his head. He was removed in an unconscious condition to Dorking, where he now remains in the Cottage Hospital.

During the Second Boer War, the town of Ladysmith was besieged for 118 days. To the amazement of local inhabitants, a tethered balloon operated by the Royal Engineers was regularly sent up to observe the insurgents' positions. The siege was finally broken on 28th February 1900 by a force under the command of General Sir Redvers Buller VC, GCB, GCMG.

**The Dorking & Leatherhead Advertiser
Saturday 14th June 1902**
HOLMWOOD

HOME FROM THE WAR. – Holmwood gave a hearty welcome, on Saturday last, to Lieut.-Col. F.C. Heath R.E., and Major G. M. Heath, R.E., on their return from South Africa. A large gathering of enthusiastic residents, both of the Holmwood and Coldharbour, assembled at the station at five o'clock, and as the gallant soldiers alighted from the train shortly after that hour they were vociferously cheered; even the torrents of rain which were falling could not quench the ardour of the assembled crowd. At Anstie Farm the carriage was unhorsed, and, amid ringing cheers, was speedily dragged to the Grange, where Sir Leopold and Lady Heath and other members of the family joyfully welcomed the returning heroes. There was some speechmaking, at any rate, in so far as the unfavourable elements would permit, followed by immense cheering and the singing of patriotic songs, which could be heard for miles around. The two officers showed the unmistakable effects of their hard campaigning. Major Heath was "the man in the balloon" during the siege of Ladysmith, whom the Boers would dearly have liked to have got at. Admiral Sir Leopold and Lady Heath tender their most grateful thanks to the very large number of friends and neighbours from South Holmwood and Coldharbour, who, notwithstanding the tempestuous weather, assembled at the Holmwood Station on Saturday last to welcome and do honour to their soldier sons.

**Dorking & Leatherhead Advertiser
Saturday 18th November 1905**
HOLMWOOD

NAVAL ABSENTEES. – Frank Jenner and Isaac Buckland, the latter a native of Holmwood, were charged before Mr. J. Carr Saunders and Mr. L. Pledge, at the Dorking Police Court on Monday, with being absentees from the Naval Barracks, Portsmouth. – P.S. Fletcher deposed to seeing both men at Holmwood station on Sunday night, where, in consequence of their failing to produce passes, he took them into custody. – Prisoners were ordered to be taken back to their barracks.

**The Dorking & Leatherhead Advertiser
Saturday 11th July 1903**
NEWDIGATE

THE FATAL ACCIDENT TO MR. ROWLAND. - An inquest was held at the Six Bells on Monday, by Mr. Gilbert White, deputy coroner, on the body of Mr. Henry Rowland, who succumbed to the injuries he sustained in the accident at Holmwood Station on June 17. From the evidence of Leonard Gardener, it appeared that on the day named deceased was at the station with his pony and cart, and as he was about to get up into the vehicle a train passed through. This startled the horse, as a result of which Mr. Rowland was thrown down, the wheel running over him, and inflicting a fractured jaw and other injuries, death ensuing last Thursday.- Dr. Young, who has attended the deceased since the accident, gave evidence as to the injuries, and the jury returned a verdict of "Accidental death".

Dorking & Leatherhead Advertiser
Saturday 25th May 1907

WHITSUNTIDE AT DORKING. – Although fine, the weather at Whitsuntide was cold and cheerless, and formed a striking contrast to the conditions that prevailed at Easter. This was a sufficient reason in itself for the falling-off in the number of visitors to the town. Still, Dorking and its immediate neighbourhood attracted, as usual, many of the holiday folk, while a steady stream of motorists and cyclists passed through the town from early morn till late at night. Leith Hill and Box Hill were thickly dotted with people, and it would seem that these favourite heights of the Surrey Hills, which command such a beautiful panorama of scenery, are becoming more and more popular with those who wish to be far from the madding crowd. On Whit Monday about 3,000 persons, a decrease on last year, detrained at Boxhill (L.B. and S.C.R.), and at Dorking on the same Company's line there were approximately 2,000. In addition to the ordinary service, 18 special trains ran to the last named station, while four through specials stopped to set down passengers. It may be mentioned that about 3,000 returned from Dorking station in the evening, the difference being accounted for by large numbers who arrived at neighbouring stations, and walked into the town later in the day. To cope with the traffic, which, by the way was successfully dealt with, six special trains started from Dorking and three from Holmwood, while five through specials stopped. About a hundred residents accompanied the Town Band by train to Crawley. Fortunately, no accidents of a serious nature have come to our notice during the holiday.

Left: A Railway Letter Service stamp issued by the LB&SCR.

In 1891, a number of railway companies made an agreement with the Post Office that letters could be handed in at a station and expeditiously sent on their way for the payment of an additional penny on the usual postage rate. Each railway company produced their own stamps, to an agreed design. This was a particularly helpful service for commercial travellers when passing on orders and for other businessmen needing to contact their offices.

The Dorking & Leatherhead Advertiser
Saturday 3rd October 1903
COLDHARBOUR

WORKMAN'S OUTING. – On Wednesday, September 23rd, the employees on the Broome Hall Estate, and their wives, between 90 and 100, were sent by Sir A. And Lady Hargreaves Brown to Portsmouth and the Isle of Wight for a day's outing. The early train arrangements were excellent, but the voyage around the Isle of Wight had to be abandoned on account of the dense fog which prevailed most of the day. Some of the more daring passengers got as far as Ryde, where they remained on board, moored to Ryde Pier all night, and only reached Portsmouth the following morning, whence they returned to their several destinations by the early morning trains. Fortunately few of the families ventured on the sea, and they arrived at Holmwood Station in excellent time the same night. Before leaving the station for their homes, the party gave three ringing cheers for Sir Alexander and Lady Hargreaves Brown for their kindness and generosity in continuing this holiday, which, with the estate cricket match, has been observed, with very rare intermission, for nearly forty years. All the excursionists enjoyed the day immensely.

Dorking & Leatherhead Advertiser,
Saturday 3rd October 1908

SUDDEN DEATH. – Miss Barbara Rose, residing at 23, Lincoln-road, Dorking, died suddenly at Holmwood Station on Wednesday. She had been attending a harvest thanksgiving service at the chapel with some friends, and was taken ill on the way to the station subsequently. P.C. Milton happened to be near, and rendered all the assistance possible. Deceased was removed to the Station, where she was seen by Dr. Robertson, but by that time life was extinct. The body was taken to Breakspear farm, to await the coroner's instructions. Deceased was about 62 years of age.

CYCLING ACCIDENT. – Mr James Adams, a porter-signalman at Ockley station was cycling near Holmwood station on Friday evening, when he collided with Mr. W.F.J. Brown, of Bregsell's farm. The latter was walking and was knocked down, but fortunately was not much hurt. Adams, however, fell injuring his arm and leg, cutting his face, and causing slight concussion of the brain. As a result of his injuries he has been unable to attend to his duties.

> **Dorking & Leatherhead Advertiser**
> **Saturday 23rd July 1910**
> *CAPEL*
>
> PARISH COUNCIL – The quarterly meeting of the Capel Parish Council was held on Tuesday evening, when there were present Mr. J F O'Byrne (in the chair), and Messrs J Worsfold, J Reeves, W H Fowler, W F J Brown, W H Geary, and Mr A J Lipscomb, with Mr E Moore (clerk). A letter from the Brighton Railway Company was read declining to stop the train leaving Dorking at 7.19 am for Horsham at Holmwood and Ockley stations. Several members of the Council expressed great disappointment, and it was resolved to invite the Ockley and Newdigate Parish Councils to join in renewing the request next year. ...
>
> ... As an outcome of the recent fatal motor car accident at Beare Green, Mr Brown called attention to the dangerous pace at which motors are driven down the hill from Holmwood station, and predicted a serious accident shortly unless some action were taken. After considerable discussion it was resolved to ask the District Council to have a warning board erected at the top of the hill.

Unfortunately, on both counts, the pleas of the Parish council seemingly fell upon deaf ears:

> **The Dorking & Leatherhead Advertiser,**
> **Saturday 21st January 1911**
> *CAPEL*
>
> PARISH COUNCIL. - ... A letter was read from the Surrey County Council, in answer to a request from the Parish Council that a motor danger sign should be erected near Holmwood Station, stating that they only erected triangles, painted red, at corners and dangerous crossings. Any special sign, with reading [sic], must be paid for by local councils. It was resolved that the County Council be asked to have a red triangle erected.
>
> ... ... Mr. Reeves proposed that the Council should repeat its request to the Brighton Railway Co, to run an earlier down train on week-days, calling at Holmwood and Ockley Stations. Mr. Geary seconded, and it was carried unanimously. It was also resolved to send a copy of the resolution to the Parish Councils of Ockley and Newdigate, and to the Dorking Rural Parish Council, inviting their co-operation.

Station Hill

Above: The view up the hill, north-westwards towards South Holmwood village. This postcard dates from 1909.

Below: A view looking the other way, towards the 'White Hart' public house and Beare Green, c.1905-10. A horse-drawn trap, carrying an unusual load, and the presence of the photographer, have drawn an audience of inquisitive local children. Note that a new canopy has been added over the front entrance to the station building.

In the summer of 1911, to celebrate the coronation of King George V, a 'Festival of Empire' opened on 12th May at the Crystal Palace. This extravaganza included the erection of ¾-sized replicas of the Parliament buildings of the countries in the Empire, in which were displayed examples of their diverse cultures and products. There was also a daily pageant celebrating the "magnificence, glory and honour of the Empire and the Mother Country". The music for this had been written by 20 composers, including Ralph Vaughan Williams and Gustav Holst, and was performed by a military band of 50 players and a chorus of 500 voices. The Primrose League organised a special train from Holmwood to take its members and their guests to visit this event:

The Dorking & Leatherhead Advertiser, Saturday 8th July 1911
PRIMROSE LEAGUE, DORKING HABITATION.
A Cheap Excursion has been arranged to the

FESTIVAL OF EMPIRE
at the CRYSTAL PALACE

from Dorking L.B.& S.C.R on
WEDNESDAY, July 12th, 1911.
The Special Train will start from
Holmwood Station at 9.55,
Dorking at 10 a.m.
Returning from Penge Station at 9.50 p.m.

Return Fare, including Admission to Exhibition:-
From Dorking:
Members, 2s 6d; Non-Members, 3s.3d.; Children under 12, 1s.8d.
From Holmwood Station:
Members, 3s; Non-Members, 3s.9d.; Children under 12, 2s.
Tickets must be purchased on or before Saturday, July 8.

MEMBERS TO WEAR THEIR BADGES

Two examples of the many Primrose League badges from the period

The Primrose League, founded in 1883, was an organisation for spreading Conservative principles in Great Britain and was named after the favourite flower of Benjamin Disraeli, the erstwhile Prime Minister. It is thought that the League enjoyed more popular support during the early years of the 20th century than the Trade Union movement and, in 1910, the membership comprised of 87,235 'Knights', 80,038 'Dames' and 1,885,746 'Associates', in a total of 2,645 'Habitations'. The Primrose League was the first political organisation to give women the same status and responsibilities as men.

From a railway operational perspective, the choice of Holmwood might seem a odd starting point for this excursion. However, as one of the principal organisers was Colonel Henry Helsham-Jones, aided by fellow Holmwood residents Mrs St. John Hornby and Mrs F S Phillips, it is highly likely that the LB&SCR was keen to maintain favour with its wealthier clientele. The 73-year old Colonel was well-known to the station staff at Holmwood for his eccentric habit of not stepping into a train until the guard had actually blown his whistle - a behaviour that may well have caused some consternation at Penge.

**The Dorking & Leatherhead Advertiser,
Saturday 7th December 1912**
FOR SALE.
Roan Cob. 14.3, six years old, quiet to ride and drive, pass all road nuisances: warranted sound; price £32. Apply - White Hart, Holmwood Station.

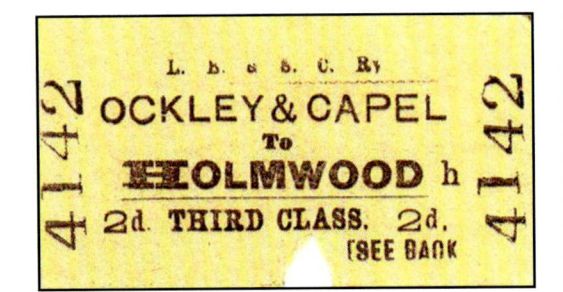

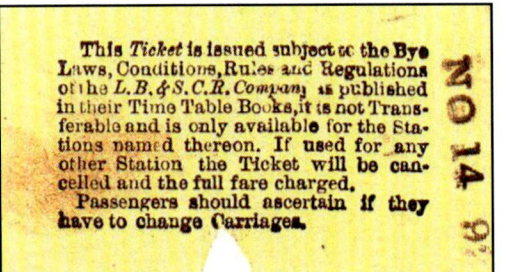

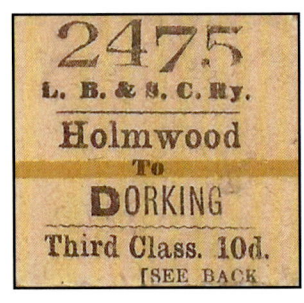

1892 single

1895 return

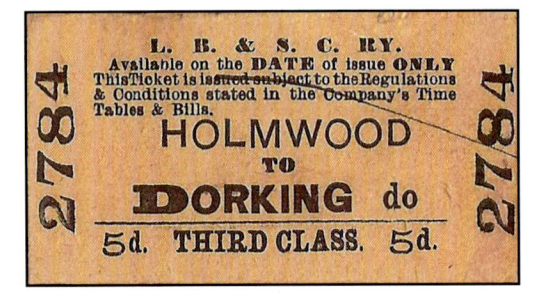

1898 return

1903 return

1916 single

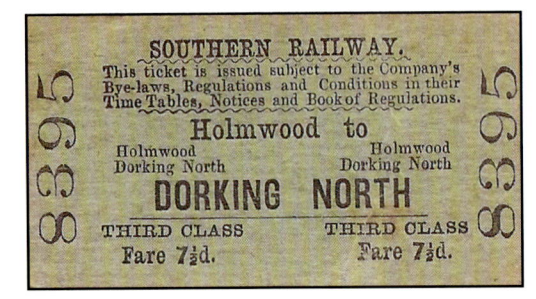

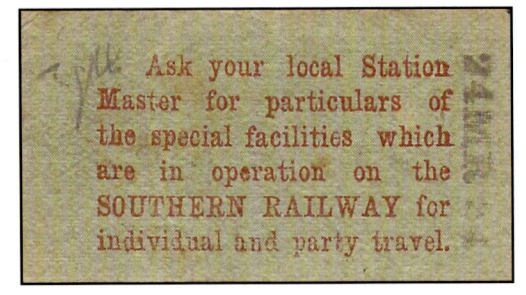

1934 single

Platform Ticket, Issued 1950

Printed Ephemera - 2

London Brighton & South Coast Railway and Southern Railway tickets
issued to and from Holmwood station,
and a Holmwood platform ticket.

62

Throughout this period, Cuthbert Eden Heath, the third son of Admiral Sir Leopold Heath, had been steadily making his mark as an underwriter at Lloyd's, specialising in novel forms of non-marine insurance - and all the while travelling up to the City from his country home, Kitlands, and then later Anstie Grange, by train from Holmwood station.

He was a pioneer in household burglary insurance and had introduced an effective method of covering diamond dealers for any losses incurred during the transportation of their wares. At the time of the San Francisco earthquake in 1906, Mr. Heath happened to be a leading earthquake underwriter and his actions after that disaster cemented Lloyd's reputation in the US. He insisted on paying all claims in relation to the earthquake and fires by famously instructing his agent in California to "pay all of our policyholders in full, irrespective of the terms of their policies" and not to quibble over the small print. As a result, Lloyd's paid out more than $50 million in claims, thought to be the equivalent of some $1billion in 2013.

His innovative approach to the business not only made him a handsome fortune, it also effectively made Mr Heath the founder of the modern Lloyd's of London insurance market.

After the Great War Mr Heath became a joint-master of the Surrey Union Hunt, whilst his summers were often spent in his holiday home, 'La Domaine de Savaric' near Eze in the Alpes-Maritime Department of France. There, much time was spent sailing in the Mediterranean on his steam yacht, 'Anne of Anstie'. This vessel was in the capable hands of a retired senior Royal Navy officer, Admiral Candy.

Cuthbert Eden Heath Esq.

Chapter 14
MORE ROYAL TRAINS AND OTHER SPECIALS

For a variety of reasons, Royal trains were often to be seen passing through Holmwood in the years leading up to the Great War. As typical examples, the race meetings at Goodwood were a regular attraction for King Edward VII, whilst Portsmouth, with its naval connections, was a destination for his successor, King George V.

A small, and probably incomplete, collection of special traffic notices setting out the operational arrangements for some of these trains in the years before the Great War still exists in the public domain, housed in The National Archives at Kew. The salient details concerning these various workings, including the passing times through Holmwood, are summarised in the table opposite.

An unidentified down LB&SCR Royal train, climbing Holmwood bank at some time prior to the Great War.

The locomotive is a class H1 4-4-2, No.39. It was designed by Douglas Earle Marsh, built by Kitson & Co., Leeds and entered service during January 1906.

The leaves on the trees and the verdant hedges show the photograph was taken in high summer, so this may be a train bound for Singleton and a race meeting at Goodwood.

The head code, "Two White Boards, with a crown painted on them, one on each end of the Buffer Beam", matches the usual specification for the LBSCR Royal Train carrying the monarch, although it was also used on at least one occasion for a train taking Queen Mary to Portsmouth Dockyard.

Date	Train	Royal Passengers	From	Departure	Destination	Holmwood
Thursday, 21st July 1910	The Royal Train, complete.	H.M. King George V, H.M. Queen Mary, H.R.H. Princess Mary, H.R.H. Prince George, and suites.	Victoria No. 7 Platform, South Station	3-00 p.m.	Portsmouth Dockyard [South Railway Jetty]	3-45 p.m. [pass]
Saturday, 11th November 1911 Return Journey	The Royal Train, complete, with Bogie Brakes Nos.155 & 442 at rear. The Royal Train, complete.	H.M. King George V, H.M. Queen Mary, H.M. Queen Alexandra, H.R.H. Princess Victoria, and suite. H.M. Queen Alexandra, H.R.H. Princess Victoria, and suite.	Victoria No. 7 Platform, South Station Portsmouth Dockyard [South Railway Jetty]	10-30 a.m. 2-50 p.m.	Portsmouth Dockyard [South Railway Jetty] Victoria, No. 7 Platform, South Station	11-15 a.m. [pass] 4-07 p.m.
Saturday, 24th February 1912 Return Journey	The Royal Train, complete. + Bogie Brake No.155, at rear.	H.M. King George V, H.M. Queen Mary, H.M. Queen Alexandra, H.R.H. Princess Victoria, and suite. ~ ditto ~	Victoria No. 7 Platform, Portsmouth Dockyard	10-00 a.m. 12.50 a.m.	Portsmouth Dockyard [South Railway Jetty] Victoria, No. 7 Platform.	10.45 a.m. [pass] 2.07 p.m.
Saturday, 3rd August 1912	The Royal Train complete, with Bogie Brake No.363 at rear, to be attached if needed.	H.M. Queen Mary, H.R.H. The Prince of Wales, H.R.H Princess Mary, and suite.	Victoria No. 7 Platform, South Station	10-30 a.m.	Portsmouth Dockyard [South Railway Jetty]	11-15 a.m. [pass]
Saturday, 10th August 1912	The Royal Train, complete, with Bogie Brake Van for luggage at the rear.	H.M. King George V, H.M. Queen Mary, and suite.	Portsmouth Dockyard [South Railway Jetty]	2-00 p.m.	Victoria No. 7 Platform, South Station	3-17 p.m. [pass]
Wednesday, 5h February 1913 Return Journey	The Royal Train, complete. ~ ditto ~	H.M. King George V, and suite. ~ ditto ~	Victoria, No. 7 Platform. Portsmouth Dockyard	10-00 a.m. 4-00 p.m.	Portsmouth Dockyard Victoria, No. 7 Platform	10-45 a.m. [pass] 5-17 p.m.
Monday, 28th July 1913	The Royal Train complete, with Bogie Brake Van No.155 at rear, to be attached if needed.	H.M. King George V, and suite.	Victoria No. 7 Platform, South Station	5-00 p.m.	Chichester	5-42 p.m. [pass]
Saturday, 2nd August 1913	The Royal Train complete, with Bogie Brake Van No.155 at rear, to be attached if needed	H.M. Queen Mary, H.R.H. Princess Mary, and suite.	Victoria No. 7 Platform, South Station	10-45 a.m.	Portsmouth Dockyard [South Railway Jetty]	11-30 a.m. [pass]
Monday, 11th August 1913	A special rake of vehicles.*	H.M. King George V, H.M. Queen Mary, and suite.	Portsmouth Dockyard [South Railway Jetty]	10-00 a.m.	Victoria No. 7 Platform.	11-17 a.m. [pass]

* "This train will consist of the following Vehicles placed in the order named on leaving Portsmouth Dockyard (South Railway Jetty):– 6-Wheel Brake Van No.380, 1st Class Bogie Lavatory Carriage No.138, Bogie Brake Van No.566, Saloon No.564, Royal Saloon No.562, Saloon No.563, Dynamo Brake No.565 and Bogie Brake Van No.155."

On Saturday, 4th November 1911 a mysterious "Important Private Special Train" was run from No.7 Platform at Victoria to South Railway Jetty in Portsmouth, and back. It was comprised of only three vehicles, Brake Van No.565, Saloon No.562 and Brake Van No.566. Its head code, a white board over each buffer and single white board with a black cross at the top of the smokebox and in the centre of the buffer beam, does not suggest a Royal use. It departed from London at 11-25 am, passed through Holmwood at 12-10, and arrived in Portsmouth at 1-5 pm. The return journey started at 4-40 pm, passing Holmwood at 5-57 and the train was back in the metropolis at 6-40 pm. But whatever was its purpose?

Less of a puzzle was the special train that took the newly elected President of France, M. Raymond Poincaré, and H.R.H The Prince of Wales, together with their respective suites, from Portsmouth Dockyard [departing at 1-30 pm] to Victoria [arriving at 3-30 pm] on Tuesday, 24th June 1913. This train, drawn by a rather gaudily decorated locomotive, passed through Holmwood station at 2-47 pm. Unusually, the Special Traffic Notice carried the unequivocal instruction, in bold type and underlined, that "It is important that the Royal Train should arrive at Victoria at 3.30 p.m. precisely, as shown in the above Time Table": the King, together with the Duke of Connaught, were waiting.

A period post card shows an artist's impression of Class H1 4-4-2 No.39. The umber-coloured locomotive has been decorated in a typical LB&SCR style, complete with a floral display beneath the smoke box and white-washed coal in its tender, to mark the visit of the French President in June 1913. It was also given the name, 'La France', in honour of this State Visit. The Special Traffic Notice stated that the crosses on the white boards of the head code should be painted red.

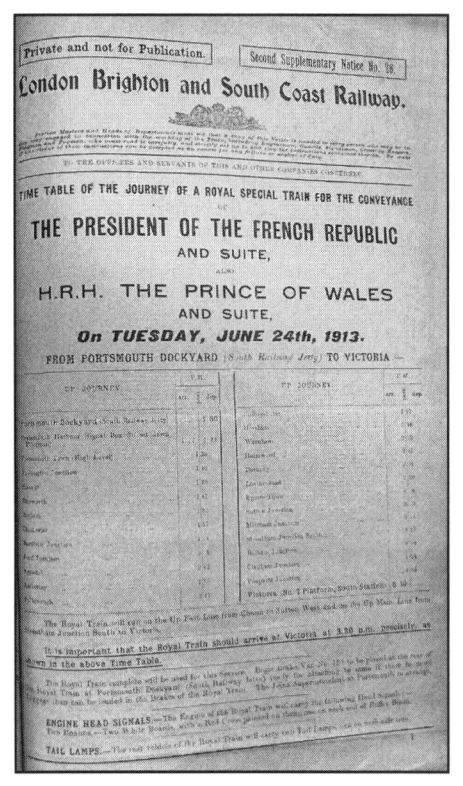

The front page of the Special Traffic Notice itself.

Apart from this being a truly splendid portrait of D1 Class No. 229 'Dorking' [built in 1884] standing outside the engine shed at Dorking, the only real justification for including this photograph is the likelihood that this locomotive visited Holmwood station as it worked between Dorking and Horsham. Its golden ochre livery, which must have looked magnificent in the sunshine, indicates that the photograph was taken before the introduction of the umber brown livery for LBSCR locomotives after 1905. A cryptic note on the back of the print says "Presented by Mr Sadler. His father is on the footplate". The LBSCR employee records show that a Mr. C. Sadler was promoted from cleaner to fireman at Dorking in July 1904, at a new pay rate of 3/- a day. Promotion to driver came in September 1919 at Epsom shed, followed by a move back to Dorking in June 1921 on 13/- per day.

Chapter 15

SUFFRAGETTES AND A SOCIAL CONSCIENCE

In 1901, the recently married Mr and Mrs Pethick Lawrence made 'The Mascot', a Lutyens designed house set in 8 acres and located a mile or so north of Holmwood station, their new home in the country. Over the next two decades it was to become a focus for various social campaigns and especially the advance of the women's suffrage movement.

Before her marriage, Emmeline Pethick had been a social reformer and co-founder of Maison Espérance (from the French, house of hope, trust or expectation), a dress-making cooperative with a minimum wage, an eight-hour day and a holiday scheme. She also formed the associated Espérance Club, which sought to improve the lot of the young women employed in the sweat-shops of the London dress-making trade, often through the medium of dance and drama.

Old Etonian Frederick Lawrence had obtained a double first at Cambridge and was called to the bar by the Inner Temple in 1899. He was independently wealthy and engaged in philanthropic work in the slums of East London when he first met Emmeline. Although they maintained an apartment in Clement's Inn, off the Strand, they entered fully into village life and, even more to the point, their London circle of friends and acquaintances did too, coming to visit them by train at week-ends. The Espérance girls were frequent visitors, as were fellow socialists and sympathetic politicians, including Keir Hardy and Ramsay MacDonald. In 1904 they had 'Sundial House' built as an independent respite home for impoverished women and children from deprived areas all over the country, a use that lasted until 1920:

> **The Yorkshire Evening Post, Monday 3rd November 1913**
> *IRISH CHILDREN IN SURREY.*
> The Irish children who were brought over to England in the midst of the excitement in Dublin, and taken down to Holmwood, Surrey, are revelling in their new conditions, not the least of which in their opinion is the luxury of the daily warm tub. They occupy Mr. Pethick-Lawrence's well-known cottage, "The Sun-dial". Lord Ashbourne allows them the run of his grounds, and they attend mass at his chapel. In addition to the ladies who take the duty of looking after their welfare, they have the mother of one of their number to keep them from feeling homesick.

In 1906, Kier Hardy introduced the Pethick Lawrences to Emmeline Pankhurst, a leading campaigner for women's suffrage and one of the founders of the Women's Social and Political Union. Whilst Emmeline Pethick Lawrence was able to join the WSPU, Frederick, being merely a man, was not. However that did not stop him using his legal training to represent the WSPU militants in court or his personal fortune to further their cause.

In 1907 the Pethick Lawrences started and financed the journal, 'Votes for Women', which maintained a circulation of 30,000 copies during the years 1908-09. Whilst their property in Clement's Inn became the London office of the WSPU, it was 'The Mascot' that was considered by the 'Daily News' to be its unofficial headquarters. It was at 'The Mascot' that suffragettes would recover from their various incarcerations and episodes of forced-feeding or practise their oratory in the garden. In the words of Annie Kenney, a former mill-worker, born in Yorkshire, and the only working class member of the WSPU elite:

Contemporary postcards showing houses in Holmwood used by the women's suffrage movement:
Left: 'The Mascot'. The photographer is standing in Mill Road and the turnpike runs across the front of the house. Holmwood station is about a mile away, to the left.
Right: 'Sun Dial House', located a few hundred yards north of 'The Mascot'. The wording above and to the left of the sundial reads, "Let others tell of storms and showers, I tell of sunny morning hours".

"Processions, Albert Hall meetings, raids on Parliament, tactics in prison, the varied forms of advertisement ... all were decided, debated, discussed, analysed and counter discussed around round the breakfast, lunch and dinner table at the Lawrence's home, in the old courts around the Strand, round the fire at Holmwood or in the woods around Leith Hill. If the beautiful woods there could have spoken, Scotland Yard would have forestalled many a militant action."

In 1908, the redoubtable Mrs Pankhurst and her daughter, Christabel, were recorded by the local newspaper as being visitors to 'The Mascot':

> **The Dorking & Leatherhead Advertiser**
> **Saturday 26th December 1908**
> *WHAT WE HEAR.*
> That Mrs. Pankhurst and Miss. Christabel Pankhurst, the two suffragettes, visited Mr. and Mrs. Pethick Lawrence at Holmwood, on their release from prison on Saturday evening.

The unusual circumstances of the Pankhursts' release from Holloway Prison provide an interesting insight into the way this struggle was being conducted. Christabel's sentence was due to expire on 22nd December, whilst her mother, Emmeline, together with another suffragette, Mary Leigh, were expected to be released on 9th January 1909. Mrs Pethick Lawrence had planned a victory procession to escort the newly released prisoners from Holloway to Clement's Inn, full details of which had been published in 'Votes for Women'.

Quite unexpectedly, the authorities released all three women on Saturday, 19th December 1908. The order from the Home Office was issued shortly before six o'clock in the evening and it was into a cold winter night that they were turned out, with no-one to greet them. Whilst Mary Leigh was able to make her way home to Camden Town reasonably easily, Mrs Pankhurst and Christabel knew that the Clement's Inn premises would be shut up and locked for the week-end. So they adopted the most sensible option open to them - they took the train to Holmwood.

70

In 1912 the WSPU organised a new campaign: one that involved the large-scale smashing of shop-windows. The Pethick Lawrences both disagreed with this strategy but their objections were ignored, especially by Christabel Pankhurst. As soon as this new line of attack began, the Government ordered the arrest of the leaders of the WSPU. Whilst Christabel escaped to France, Frederick and Emmeline Pethick Lawrence, together with Emmeline Pankhurst, were arrested and charged with "conspiracy to incite persons to commit malicious damage". Even though the Pethick Lawrences had not carried out any of the vandalism, they were found guilty and sentenced to nine months imprisonment. Whilst in gaol they both went on hunger strike and were forcibly fed.

But that was not all. The personal wealth of Frederick Pethick Lawrence meant that he was a prime target not only for civil actions for damages in respect of the broken windows, but also the Government, yet again, who sought to recover the costs of his recent prosecution under new powers against those providing financial assistance to the WSPU.

On their return from Canada, where they had been recuperating from the effects of their prison experiences, the Pethick Lawrences found that bailiffs were in possession of 'The Mascot' and its entire contents were being advertised for sale in the Dorking Advertiser and elsewhere.

Far Left: A post card portrait of Mrs Emmeline Pethick Lawrence, c.1908.

Top Left: Miss Christabel Pankhurst, 1907/8. This dramatic pose has been struck in front of a blanket strung up in a back-yard and not whilst addressing political meeting

Top Right: Mrs Emmeline Pankhurst starts to get down from a London & South-Western Railway carriage at Waterloo station, c.1910. This is also a posed photograph.

Bottom Left: Miss Annie Kenney, 1909.

Bottom Right: Mr Frederick Pethick Lawrence, c. 1910.

Each of these portraits shows its subject before the ravages of imprisonment and forced-feeding really took its toll on them all.

AUCTIONEERS' ANNOUNCEMENTS.

MR. G. M. FRIEAKE.

The Director of Public Prosecutions v. F. W. Pethick Lawrence.

"THE MASCOT,"
HOLMWOOD, near DORKING, SURREY.

G. M. FRIEAKE

will Sell by Auction, as above, on THURSDAY, 31st OCTOBER, 1912, at One o'clock precisely, the CONTENTS OF THE ABOVE WELL-FURNISHED RESIDENCE, in Early English style, including Oak and Enamelled Bedroom Suites, ditto Bedsteads, Dining and Drawing-room Furniture, Settees, Lounge and Easy Chairs, Full-compass Piano by Bechstein, small Library of Books, Pictures, Wines and Spirits, a quantity of Linen, Plate, China, and Glass, Copper Culinary Utensils, Outdoor and numerous other effects.

May be viewed day prior and morning of Sale. Catalogues 89, Chancery-lane, W.C.

**Dorking & Leatherhead Advertiser
Saturday 2nd November 1912**
SUFFRAGETTE SALE
THE GOVERNMENT AND MR. AND MRS. PETHICK LAWRENCE

The long anticipated sale of the contents of "The Mascot", the Holmwood residence of Mr. and Mrs. Pethick Lawrence, took place on Thursday. It will be remembered that at the Suffragette conspiracy trial in May last Mr. and Mrs. Pethick Lawrence, with Mrs Pankhurst, were not only sentenced to a term of imprisonment in the second division, but were ordered to pay the costs of the prosecution, variously estimated up to £900. The amount not being forthcoming, the Government issued a distraint, and for several weeks past the bailiffs have been in possession of "The Mascot". Rumours of possible trouble have been very general during the past few days, and as a result there was a large attendance at Thursday's proceedings - probably from three to four hundred. The Suffragettes, too, were out for support. Large posters announcing the sale appealed to the public to support Mr. and Mrs. Pethick Lawrence in "their fight against tyranny by their presence and sympathy ... ...".

The police were prepared for any emergency. Under Supt. Coleman there was a force of 20 present, and as the constables were for the most part in mufti, they were hardly noticed. The members of the Women's Social and Political Union, as well as of the Women's Tax Resisters' League, were strongly in evidence; the former were easily recognised by their colours – purple, white and green, and it was noticed that many wore their badge of imprisonment. Quite a large muster of the supporters of women's suffrage arrived from town by the train arriving at Holmwood shortly after noon, and for the remainder of the afternoon a feeling reigned of excitement and expectancy quite foreign to the usual quiet and serene atmosphere of the village. The scene in front of the house soon became a very animated one. Only a small proportion of the assembled company were able to inspect the contents of the house, even if they had the disposition to do so, which apparently few had; the large majority were content to remain in the grounds, and to await developments. But if they anticipated anything unusual they were doomed to disappointment.

A few minutes before one o'clock – the hour announced for the commencement of the sale – Mr and Mrs Pethick Lawrence arrived. They had walked up from their little cottage known as "The Sundial", where they had been receiving a few of their more intimate friends. Their reception was a very warm and cordial one, and Mr Pethick Lawrence lost no time in speaking the few words which he had promised. ... ...

Mrs. Pethick Lawrence was received with prolonged cheering and waving of hats and handkerchiefs. She thanked all her friends and neighbours for their sympathetic attendance and said the household possessions rendered dear to her husband and herself by use were now laid out for their inspection, and would shortly be knocked down to the highest bidder (shame). ... ...

The sale then proceeded, and by five o'clock the whole of the 280 odd lots were disposed of. The auctioneer was Mr. G. Frieake of 89, Chancery-lane, and before offering the first lot he expressed his appreciation of the courtesy Mr. Pethick Lawrence had shown him and his generosity to his men, to whom he had behaved in the most kindly manner.

What this article does not say is that Mr Frieake, like so many others attending the sale, bought an item and gladly returned it to Mr and Mrs Pethick Lawrence.

Above:: Class I3 4-4-2 tank engine No. 22 breasts the summit of Holmwood bank with a train from London Victoria to Portsmouth, via Sutton, Epsom, Dorking & Horsham. Having just whistled for the accommodation crossings at Bregsells Farm, the driver will soon have the Holmwood down distant signal in sight.

Opposite: A rather more formal portrait of No.22, on a post card produced by the Locomotive Publishing Company. This locomotive first entered service in March 1908 and was finally withdrawn in May 1951. Designed by Douglas Earle Marsh, unlike some members of the class, it was fitted with a superheater from new, at the insistence of B K Field, his Chief Draughtsman. Thus it was the first express locomotive in Britain to carry this improvement, which produced notable economies in water and coal consumption, whilst providing sufficient power for the heaviest LB&SCR passenger trains. The last ten members of the class [Nos. 82 - 91] were dual-fitted with both Westinghouse and vacuum brakes, which made them particularly suitable for working ambulance trains during the Great War. On No. 22, the air pump for the Westinghouse system is prominent in front of the side-tank. These were efficient, capable and handsome locomotives.

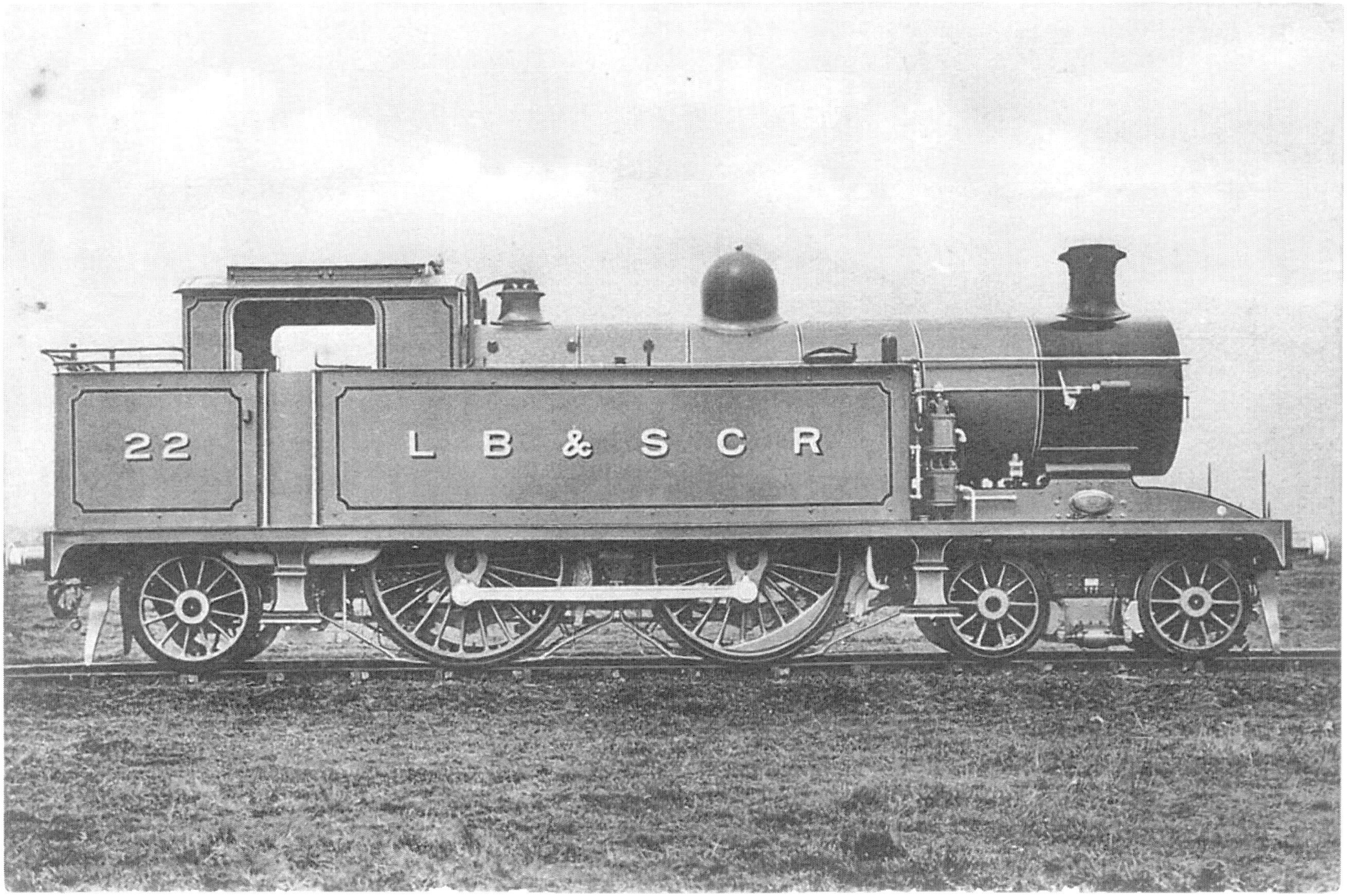
22
L B & S C R

Chapter 16
THE GREAT WAR AMBULANCE TRAINS

In 1916, at great personal expense, Mr Cuthbert Heath converted his own home, Anstie Grange, into a hospital for wounded officers. The mansion, situated on the slopes of Leith Hill, was short carriage drive away from Holmwood railway station and it readily offered the peace and tranquillity required for this new use. These selected press cuttings These selected press cuttings provide an insight into this contribution to the war effort on the Home Front:

> **Dorking & Leatherhead Advertiser**
> **Saturday 14th October 1916**
> *HOLMWOOD*
> WAR HOSPITAL FOR WOUNDED OFFICERS – Another large convoy of wounded officers was received at Anstie Grange on Tuesday evening. The wounded were met at Holmwood station by the ambulances and stretcher-bearers of the R.A.M.C., who received valuable help from members of the Local Volunteer Regiment under Mr. V. Harding, the whole work being under the supervision of Dr. McComas and Dr. Alderson. A number of local residents kindly lent motor cars for the conveyance of the wounded, who came straight from the Western Front.

> **Dorking & Leatherhead Advertiser**
> **Saturday 24th February 1917**
> *ANSTIE GRANGE OFFICERS' HOSPITAL*
> Two convoys of wounded officers, straight from the Western Front, arrived during the week at this hospital. At present this well-equipped hospital contains over 60 patients. As usual, Dr. McComas, the medical officer, was present at the railway station as the ambulance train arrived... .

> **Dorking & Leatherhead Advertiser**
> **Saturday 4th November 1916**
> *HOLMWOOD*
> ANSTIE GRANGE WAR HOSPITAL – Another convoy of wounded officers (25 cot and 10 walking cases) was received at the above on Saturday evening. The transfer from the train and removal to the hospital was effected by members of the R.A.M.C., assisted by local Volunteers under Mr. Harding. Under the care of Dr. McComas and the efficient hospital staff, all the wounded are making satisfactory progress. Although the hospital has only been open since the middle of September, over 100 patients have been received.

> **Surrey Mirror & County Post**
> **Friday 13th April 1917**
> *LATE ADVERTISEMENTS*
> WANTED, Strong Girl as Scullerymaid for Officers' Hospital: experience not necessary if willing to learn: good wages to anyone suitable. – Apply Commandant, Anstie Grange Hospital, Holmwood.

In the early spring of 1917, one ambulance train brought home one of Mr Heath's nephews. His chance arrival at Anstie Grange went some way to healing a family rift, as noted in Genesta Heath's diary.

Her account then goes on to record that Genesta's mother was "... always at daggers drawn with Fred's parents at their house, Kitlands, and they were not allowed at Anstie. But she could not forbid them visiting their son at Anstie, and they came through the wood every day to see him".

Dorking & Leatherhead Advertiser
Saturday 3rd March 1917
DORKING'S ROLL OF HONOUR
CAPT. HEATH WOUNDED.

Captain F. D. Heath, 2nd Sussex Yeomanry, attached to the 10th Queen's Royal West Surrey Regiment, only surviving son of Mr. and Mrs. Heath of Kitlands, Holmwood, was severely wounded while in a front line trench on February 10th. Capt. Heath, who received a gunshot wound through the neck, is now at Anstie Grange Hospital, and is doing well. His elder brother was killed at the battle of Loos.

Diary of Genesta Heath

[Mr C E Heath's daughter - despite her position in the household, she was working as a pantry maid in the hospital. Hence the upstairs/downstairs references.]

16th February 1917

This evening we were told that a convoy of thirty was coming in at midnight. Great excitement among the upstairs people, as there were only twenty-five free beds, but someone raised some others and everything was beautifully arranged. It was 2.30 when the convoy arrived – thirteen walking cases and seventeen stretchers. The last man was tremendously bandaged and looked – what we could see of him – very ill. We sent up drinks and waited until 3.45, when I was sent to bed because I had been up early. And the last case, the very last to be taken in, was Fred Heath himself!

17th February 1917

Commandant took me up to see Fred this morning.. ... It is extraordinary that he should be here – he had not asked to come, nor even mentioned the place. How strange things work out.

Right: An artist's impression of the various ambulance trains in service during the Great War, published in 'The Graphic' Summer Number on 16th June 1917. The accompanying text provides these details: "A standard British ambulance train has, like the French train, 16 cars, and has three medical officers, some eight nurses, many N.C.O.s and orderlies attached. It is in command of a major, R.A.M.C., and runs between casualty clearing stations at the railhead within sound of the guns and the base hospitals".

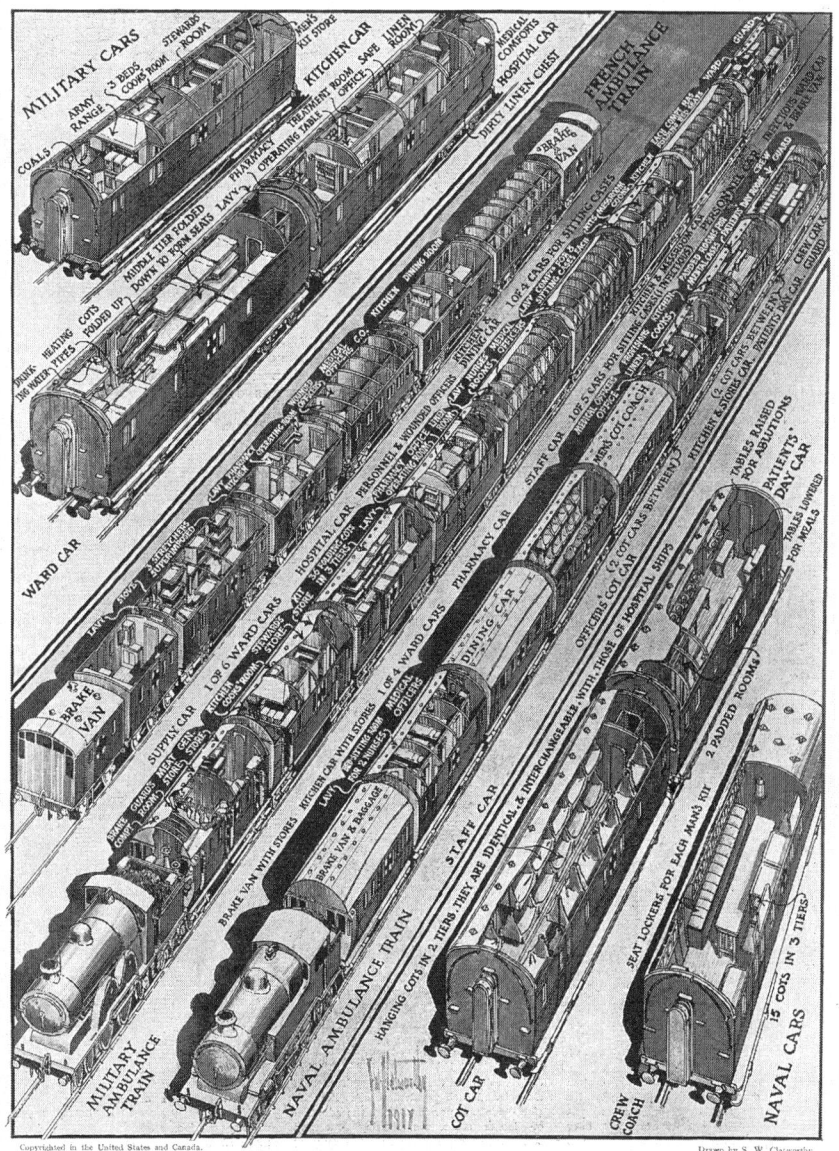

THE BEST SUMMER HOLIDAY – BACK TO BLIGHTY

Anstie Grange Military Hospital.

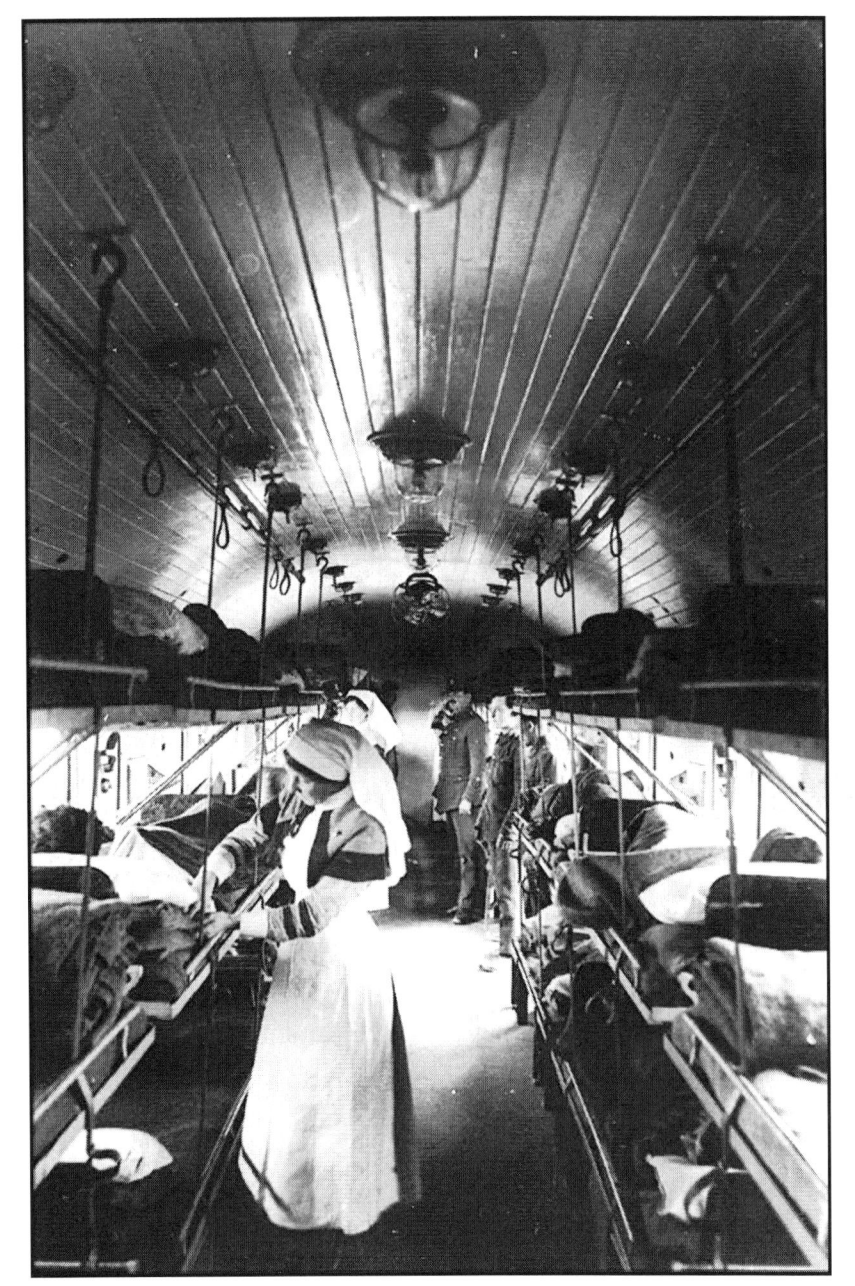

A perhaps unexpected aspect of railway operations established by this diary is that the ambulance trains arrived at Holmwood in the dead of night. The London, Brighton & South Coast Railway Ambulance Train Notice No.1 warns that "Special Ambulance Trains may be expected to run from Southampton (L&SWR), Dover (SE&CR) and Avonmouth (GWR) system to stations on the LB&SCR Company's system at uncertain times" until further notice. Amongst the strict operation instructions is a requirement that "…no unauthorised person must be allowed on a platform at which an ambulance train is being loaded or unloaded, and the issue of platform tickets to such platforms must be suspended. A public notice to that effect had to be posted accordingly".

The same notice lists the 28 rakes of vehicles supplied by the Great Central; Great Eastern; Great Western; Lancashire & Yorkshire; London & North Western; Midland; London & South Western and South East & Chatham Railway Companies, respectively [see Appendix: 6]. Whilst specific operating instructions are given for certain stations, there are none specifically for Holmwood, other than a passing reference in an instruction to "trains from Dover to Holmwood and beyond" stopping to unload at West Croydon. Bearing in mind that only the down platform at Holmwood has ready access on the level, this omission suggests that ambulance trains only ever approached Holmwood in the down direction. This point seems to be confirmed by the running times set out in the Notice, see Table Nos. 13-16 in the extract set out below.

As the 'Empty Trains' referred to in these running tables actually contained medical staff, they were to be worked home expeditiously.

Opposite Page

Left: An interior view of an ambulance railway carriage in use.

Top Right: An artist's impression on a contemporary post-card showing a 7 vehicle ambulance train running on the LB&SCR during the Great War, hauled by class J2 4-6-2 tank locomotive No.326 "Bessborough". The original design of this locomotive, by Douglas Earle Marsh, was modified by his successor, Lawson Billinton, prior to it entering service in March 1912. The locomotive was finally withdrawn from service in June 1951. The vehicles in the train are painted olive drab and carry understated international Red Cross symbols. The locomotive is in the umber livery introduced by Marsh in 1905. The head code [the white disks, with black crosses] is appropriate for a special train run in daylight on the Victoria to Portsmouth route, via Mitcham, Sutton, Epsom, Dorking and Horsham, which passes through Holmwood.

Bottom Right: A contemporary post-card showing the Anstie Grange Military Hospital. At least one of the recuperating officers playing croquet is doing so wearing a dressing gown.

This Page

Right: A lapel badge issued to its employees by the LB&SCR during the Great War. Such badges were worn by the railwaymen of all the various railway companies to show that they were in fact contributing to the war effort and were not shirkers. This was in response to the unthinking issue of white feathers and other misplaced abuse heaped upon railway workers by the, generally female, civilian population in the early years of the conflict.

Chapter 17
HOLMWOOD SIGNAL BOX

When the railway first opened, it is thought that the signals at Holmwood station were controlled from a small wooden hut on the up platform, situated about 90 yards from its Ockley end, adjacent to the waiting shelter. However rapid developments in railway safety, including the mechanical interlocking of signal and point mechanisms controlling a specific route, whilst ensuring that other signals in the vicinity were locked against improper use or indicating conflicting operating instructions, meant that the various levers used to control the various signals and points needed to be housed in a special covered and glazed structure – a signal box.

During the period from about 1860 to around 1890, the firm of Saxby & Farmer were the dominant force in railway signalling equipment manufacture and much of their equipment survived in regular railway use throughout the United Kingdom until 2006, when 'rail operating centres' started to take over this work. The patent holder of this crucial interlocking system was John Saxby, a former carpenter who had once been employed by the LB&SCR. His partner, John Stimson Farmer, had once been an assistant to the manager of that railway.

The current signal box at Holmwood dates from 1877. It is an early and complete example of a Saxby & Farmer type 5 signal box, one of their most successful and enduring designs. It is a style particularly associated with the LB&SCR, although other examples were built on more than a dozen railways elsewhere. Currently, four former LB&SCR type 5 boxes, alas all no longer in active use, are on the Statutory List of Buildings of Special Architectural or Historic Interest maintained by Historic England ('listed buildings'). They are Havant (1876); Isfield (1877); Crawley (1877) and Holmwood.

Holmwood Signal Box (1) - Photographed on 4th June 1969.

The structure stands on a plinth of red brick, laid in Flemish bond, and is of wood-framed construction. To the front it has a pair of 4-pane side-sliding sash windows, the upper lights of which have curved heads. Above them is a deep-panelled frieze, together with the brackets, with faceted finials, that support the boxed-eaves and oversailing. The hipped roof is of Welsh slate, with a central ridge vent.

Inside, it still contains the original Saxby & Farmer 18-lever frame and levers [visible through the lower window panes], although regrettably the Victorian telegraph block instruments, with their brass-cased signal and track circuit repeaters, made by the W R Sykes Interlocking Signal Company [and carried on the shelf above the levers] have been removed in recent years. The machinery room, housing the locking apparatus for signals and points, is beneath the floor of the box, below platform level.

Also of note are the enamel sign with the station name and the tail lamp.

Holmwood Signal Box (2)

THE STANDARD B.R. SIGNALBOX BELLCODE

Message	Beats on Bell
Call attention	1
Is line clear for:—	
Express Passenger train	4
Ordinary Passenger train	3—1
Branch Passenger train	1—3
Non-passenger train composed entirely of vehicles conforming to coaching stock requirements	1—3—1
Express freight train pipe-fitted throughout with automatic brake operative on not less than half the vehicles	3—1—1
Empty Coaching Stock train	2—2—1
Express freight train with automatic brake operative on not less than one third of the vehicles	5
Express freight train with automatic brake operative on not less than four vehicles, or	
Express freight train with a limited load of vehicles **not** fitted with continuous brake	1—2—2
Express freight train **not** fitted with continuous brake	3—2

Message	Beats on Bell
Light engine	2—3
Engine with one or more brake vans ...	1—1—3
Through freight or ballast train ...	1—4
Mineral or empty Wagon train ...	4—1
Freight train stopping at intermediate stations	3
Branch freight train	1—2
Freight train requiring to stop in section	2—2—3
* Train entering section	2
† Train out of section, or obstruction removed	2—1
Obstruction danger	6
Train passed without tail lamp	9 (‡) 4—5 (§)
Train divided	5—5
Stop and examine train	7
Closing Signal Box	7—5—5
Opening Signal Box	5—5—5

* Rear box warns box ahead that train has passed.
† Box ahead notifies rear box that train has arrived.
(‡) To box ahead. (§) To box in rear.

Above: The exterior of the signal box in January 1923. This photograph clearly shows the original platform level before it was raised by the Southern Railway using pre-cast concrete panels and how the 'leadaway', taking the point-rodding and the signal wires from the machinery room, was once bridged by stout timber planking. The brick-edging to the platform, and its lack of white line, should also be noted. The signalman is believed to be a Mr Grantham.

Top Left: The interior of the signal box, showing the original lever frame, the point and signal levers and the track circuit repeaters and telegraph block instruments on the shelf above.

Bottom Left: The sound of bells was ever-present in a signal box as a train was 'passed on' from the box 'in rear' and then to the box 'in advance' as it made its progress up the line.

79

This photograph, taken from the turnpike bridge looking north-east, shows Marsh class I1 4-4-2 tank engine No.603 bringing a down train into Holmwood station. The locomotive entered service in June 1907, was rebuilt as class I1X in August 1928 and finally withdrawn from service in April 1951. The headcode, used by the LB&SCR between 1910 and 1917, indicates that it is a London Bridge to Portsmouth train, via Epsom, Dorking and Horsham.

The trailing cross-over from the up to the down line was lifted in the autumn of 1927. The Holmwood up advance starting signal is just visible through the smoke from the locomotive's chimney. It was in the woodland to the right of this picture that the body of Miss Phyllis Shakespear was found in 1935.

Had the Dorking Brighton and Arundel Atmospheric Railway been built, it would have crossed this view from right to left, near the tail of the train.

The plumes of steam issuing from the side tanks of the locomotive are explained on page 82.

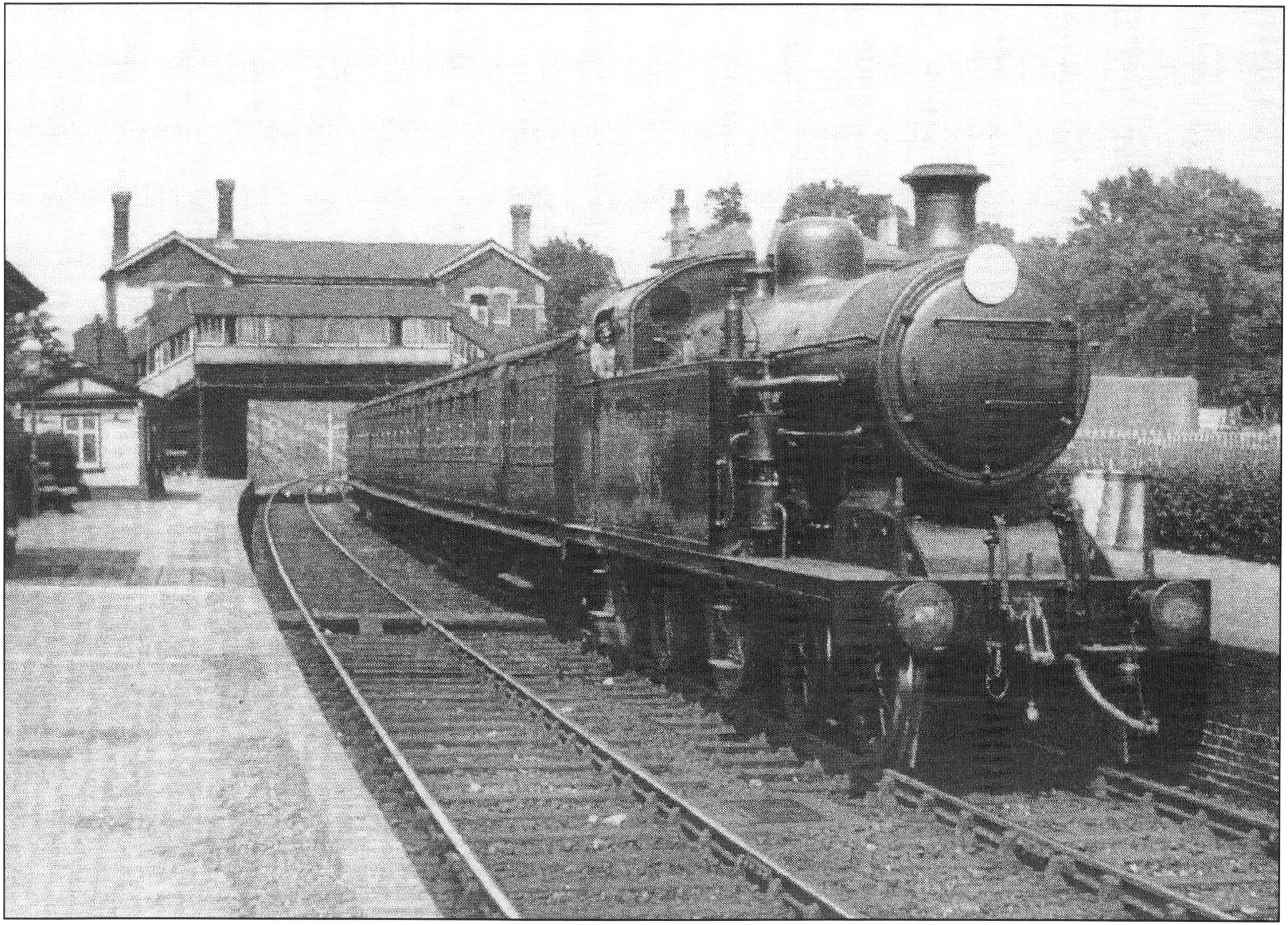

Previous Page: Following the 'grouping' of the railways in 1923, the London, Brighton & South Coast Railway was subsumed into a new conglomerate company, the Southern Railway. Although taken in 1927, this photograph shows the LBSCR architecture and infrastructure at Holmwood station in its final form, before the SR started to make changes. A covered bridge, with sliding side-glazing, takes passengers from the booking hall to the covered, open-sided, staircases that lead down to platform level. On its tall wooden post, the aspect of lower quadrant up starting signal stands out clearly against the square of white painted brickwork on the main station building. Through the bridge opening can been seen the equally tall post for the up advanced starting signal, and the adjacent trailing cross-over from the up to the down line. On the down platform, next to the locomotive, sit a group of 17 gallon milk churns. They are probably returned empties, awaiting collection by the farmer. As each churn weighed 2 hundredweight when full, they were moved about the station on a three-wheeled barrow. During this period, the station name boards carried the wording 'Holmwood for Leith Hill'.

In contrast, class I1X 4-4-2 tank engine No. B595 in has undergone a transformation whilst in Southern Railway ownership.

When first introduced by the LB&SCR in September 1906, the I1 class was designed by D E Marsh to haul secondary passenger trains, especially in the south London suburbs. The entire class of the twenty locomotives was constructed by Brighton works, the first ten being numbered 595-604 and the remainder numbered 1-10. The Southern Railway initially added a "B" prefix to these numbers and later renumbered them 2595-2604 and 2001-2010.

In polite society, because many of the class were eventually allocated to Brighton or Horsham engine sheds, they were known as 'Wealden Tanks'. However, to the footplate crews who worked on them, they were known by a much earthier epithet on account of their poor steaming capabilities and other unpleasant characteristics.

As built, the I1 class condensed their exhaust into the side tanks to pre-heat the feed-water, prior to it being pumped into the boiler by two cross-head driven pumps and an extra water pumping cylinder beneath the Westinghouse air brake pump. As the water in the side tanks got hotter and hotter, particularly when the locomotive was working hard, so did the crews.

Initially, to try and alleviate this problem, clerestory roofs were fitted to the cabs. Later on, the tanks were lagged and then eventually fitted with an internal baffle plate to prevent the hot water getting back as far as the cab. In the photograph on page 80, as a consequence of the locomotive having been worked hard up Holmwood Bank, steam is issuing from the vent pipes located either side of the front spectacle plate, whilst in the inset picture [below] such a vent pipe is clearly visible coming out of the top of the tank, above the 'B' in LB&SCR lettering on the tank-side.

To try to improve their steaming, in 1921 Lawson Billinton fitted taller chimneys to the class but it was not until they were all rebuilt by Richard Maunsell as class I1X, using the larger boilers removed from locomotives of the B4 and I3 classes, that these engines were able to match the D1 class tank engines they replaced. No.595 was rebuilt in January 1925 and it is in this form that this locomotive appears in this photograph, wearing the new SR livery of 'Maunsell Green'. This was set off with white lining and black edging. The locomotive's wheels were painted green, with black tyres and centres, whilst the buffer beams were in unlined red. The lettering and numerals are in Mid Chrome Yellow, to simulate gold leaf.

And, judging from the white feather at the safety valves, the locomotive has plenty of steam to spare!

No. 595 as originally built.

The LB&SCR had long used publicity to encourage the use of its system [below, left]. Its successor, the Southern Railway, followed suit with some iconic ideas of its own.

Below Right: A 1932 poster by Herbert Edmond Vaughan.

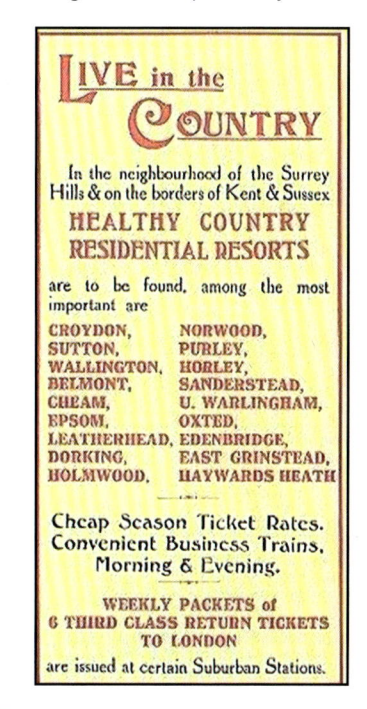

LIVE IN SURREY FREE FROM WORRY.
FREQUENT ELECTRIC TRAINS DAY AND NIGHT
"THE COUNTRY AT LONDON'S DOOR"
FREE AT ANY S.R. ENQUIRY OFFICE.

SOUTHERN ELECTRIC — THE QUICKEST WAY HOME
H. A. WALKER, GENERAL MANAGER.

LIVE in the COUNTRY
In the neighbourhood of the Surrey Hills & on the borders of Kent & Sussex
HEALTHY COUNTRY RESIDENTIAL RESORTS
are to be found, among the most important are

CROYDON, NORWOOD,
SUTTON, PURLEY,
WALLINGTON, HORLEY,
BELMONT, SANDERSTEAD,
CHEAM, U. WARLINGHAM,
EPSOM, OXTED,
LEATHERHEAD, EDENBRIDGE,
DORKING, EAST GRINSTEAD,
HOLMWOOD, HAYWARDS HEATH.

Cheap Season Ticket Rates.
Convenient Business Trains.
Morning & Evening.

WEEKLY PACKETS of 6 THIRD CLASS RETURN TICKETS TO LONDON
are issued at certain Suburban Stations.

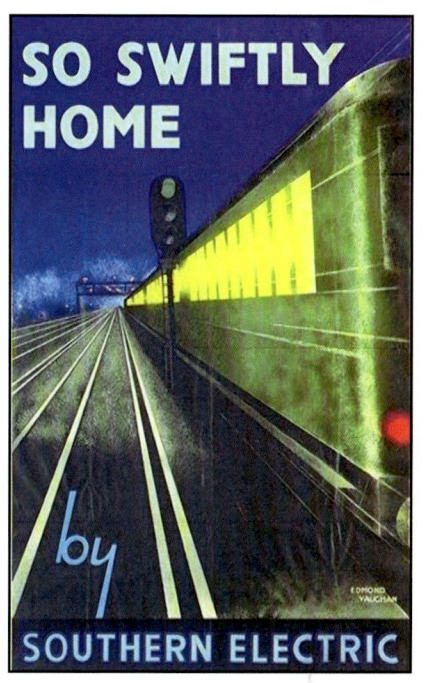

SO SWIFTLY HOME by SOUTHERN ELECTRIC

SOUTHERN ELECTRIC

Above: The 1930's 'Southern Electric' logotype used to publicise the new electric train services, with its distinctive lightening flash and sans-serif font with an inner line.

Left: A poster by Ethelbert White [1891-1972]. Its sentiment helped to boost the sale of 'homesteading' plots on the White Hart Estate adjacent to Holmwood station, especially during the Second World War.

Overleaf: A booklet published by the Southern Railway in the 1930s and a map showing a ramble that starts or finishes at Holmwood station. A 'Go-As-You-Please' Cheap Day return ticket from either Victoria or Waterloo to Holmwood cost 6/3d [1st class] and 3/9d [3rd class].

SOUTHERN RAMBLES
FOR LONDONERS
by S.P.B. MAIS

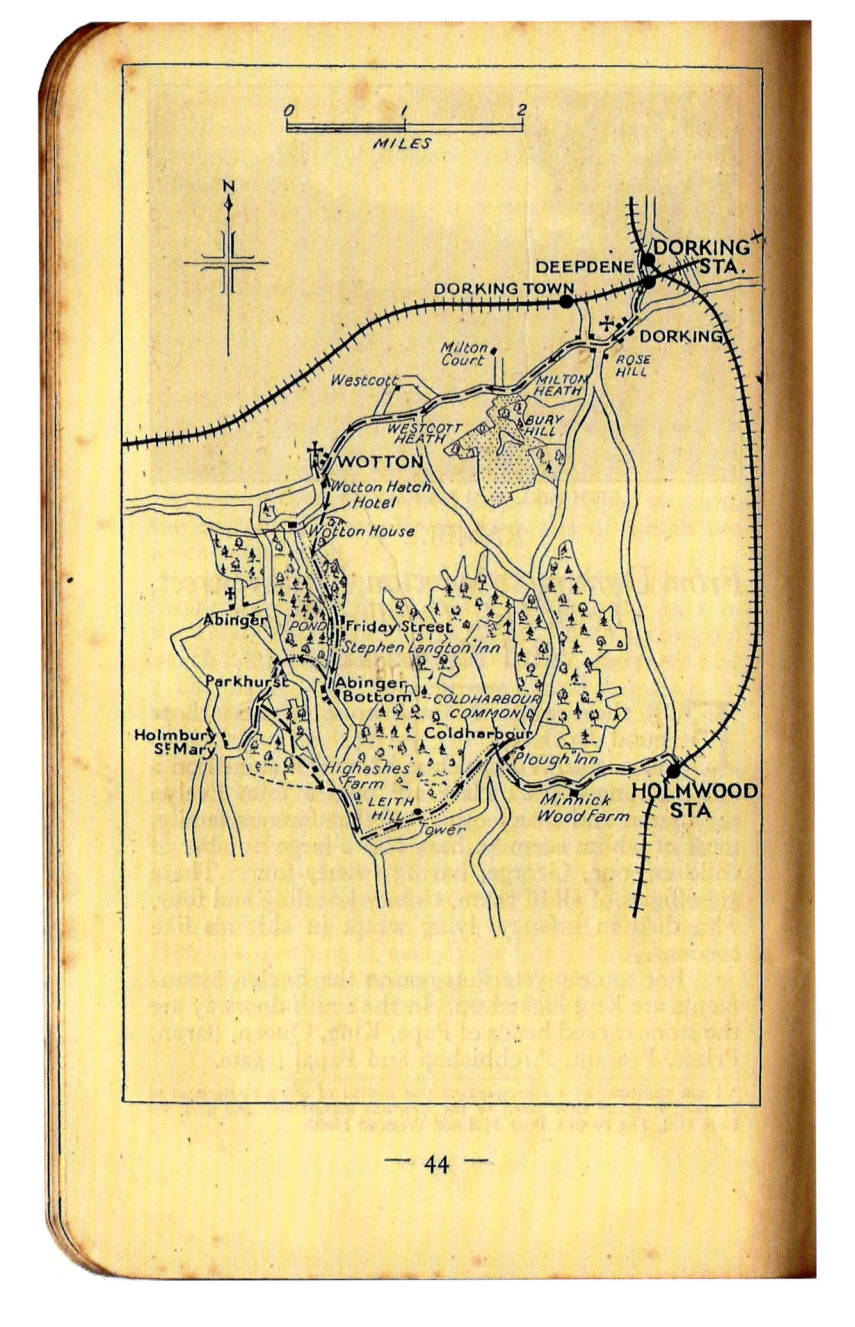

Published by the Southern Railway
PRICE SIXPENCE

— 44 —

Chapter 18
BETWEEN THE WARS

During the Great War, in order to ensure that necessary strategic overview could be taken and to remove the distracting internal completion between the rival railway companies, Parliament decided that the greater national interest would be served if the railways in Great Britain were placed under state control. This situation continued until 1921. Indeed, such was the parlous condition of the railways during this period that the post-War government, led by David Lloyd George, gave serious consideration to their complete nationalisation.

In the end, however, this proved to be a step too far and the myriad independent railway companies were subsumed into four new, larger, companies, often known as the 'Big Four'. The aim of this amalgamation was clearly set out in section 1(1) of the Railways Act of 1921:

RAILWAYS ACT 1921 [11 & 12 Geo. 5. Ch. 55.]

PART I.

REORGANISATION of RAILWAY SYSTEM.

1.(1) With a view to the reorganisation and more efficient and economical working of the railway system of Great Britain railways shall be formed into groups in accordance with the provisions of this Act, and the principal railway companies in each group shall be amalgamated, and other companies absorbed in manner provided by this Act.

This grouping took effect from 1 January 1923. Thus it was on this date that the railway line between Dorking and Horsham became part of the Southern Railway (SR). However, as the photograph on page 81 shows, apart from changes in Company nomenclature and to locomotive and rolling stock livery, there were no discernable or substantive changes at Holmwood station for several years.

Meanwhile, country life continued in its usual course and the Dorking Rural District Council got worked up about a storm in a teacup:

Surrey Mirror and County Post, Friday 21st November 1924
DORKING RURAL COUNCIL
COLDHARBOUR'S NEED

Reporting on the result of the parish meeting recently held at Coldharbour to discuss the necessity of a road between Coldharbour and the Holmwood district and station, Mr. Carter said the meeting was largely attended, And Mr. Arthur Heath, on behalf of himself and Mr. Cuthbert Heath, stated they had no objection to the provision of the road; personally, he felt it was not particularly needed but nevertheless his brother and himself were willing to grant every facility for widening the existing road. This statement by Mr. Heath went a long way to "clear the air" and gave him (Mr. Carter) the idea that a good deal of the controversy could have been avoided. Mr. Heath was followed by Mr. Lipscombe and Mr. Weller, both members of the Capel Parish Council, and representing Coldharbour.

They gave concrete instances of the narrowness of the existing road and the need which was felt in the village for a road to Holmwood station. Other persons present at the meeting followed, and all seemed to speak in the same strain, except for one man, who was apparently a stranger. The meeting appeared to be practically unanimous that something should be done either in the direction of the provision of a new road, or the improvement of the existing one.

The Chairman having thanked Mr. Carter for his report, the Committee decided to further consider the matter in committee.

It was decided eventually, to leave the matter in abeyance for the present.

> **Surrey Mirror and County Post**
> **Friday 1st September 1922**
> FOR SALE. 10 two-year old Hens, 5 Light Sussex, 3 Leghorns and 3 Cockerels – Annetts, opposite Holmwood Station

Elsewhere, rather more momentous changes on the railway were taking place. Coincidentally, these were reported in the very same newspaper as the Coldharbour controversy:

> **Surrey Mirror and County Post**
> **Friday 21st November 1924**
> *SOUTHERN RAILWAY*
> *THE WORK OF ELECTRIFICATION*
> … On the Brighton section work has been proceeding very satisfactorily on the Victoria to Coulsdon and Victoria to Sutton lines, which will have electric trains in February. On the South-Western section electrification is proceeding on the lines from Raynes Park through Leatherhead to Guildford. The section from Leatherhead to Dorking is to be electrified on the third rail system. Dorking is on the Brighton line, which employs the overhead system, and the effect of electrification upon Dorking will be to change its London terminus from Victoria to Waterloo.

Although the LB&SCR had decided to run electric trains using single phase alternating current at 6600v 25 Hz delivered by overhead wires as early as 1904, their expansion of this system was slow and further curtailed by the Great War. Responding to the same threat, that posed by the growing electricity powered trams on the road network of the Metropolis, the London and Southwestern Railway [LSWR] elected to use a system that drew its electrical power from a third rail. This carried a 660v direct current and was supported on ceramic insulating pots screwed into the outer edge of the sleepers that formed part of existing track. Thus, at the time of the grouping, there were two competing systems in South London for the operation of electric trains.

Realising the commercial advantages that electric trains offered, the newly formed Southern Railway opted for the third rail system, which was extended to Dorking with effect from 12th July 1925. From that date, a regular interval service was run to Waterloo twice an hour, seven days a week. Starting on 3rd March 1929, a similar service started to run from Dorking to London Bridge, also twice per hour at peak periods. These changes had a profound effect on the passenger train service from Holmwood.

From about 1907, the LB&SCR had started to experiment with the use of 'motor' or 'pull-push' trains between Horsham and Dorking. These saved on the need for a locomotive to have to run round its train at the end of each journey. When the engine was pulling its train, the normal arrangements applied; however, when the locomotive was propelling [pushing] the carriages, the driver could control the regulator on the engine and the brakes from another cab, located at

[See page 85] It is not clear whether the Coldharbour residents were complaining about Anstie Lane, the county road [which does not really go to Holmwood station], or Moorhurst Lane, the bridleway shown in this contemporary postcard. The latter forms part of the direct route from the station up the hill to the village and was metalled for the use of horse-drawn vehicles..

the front of the train, through an arrangement of rods and linkages. He could also communicate with the fireman, who remained on the locomotive footplate, by means of a series of bell signals. For safety reasons, no more than two carriages were propelled in a single train.

Although the through steam services from Holmwood to Victoria and London Bridge had carried on much as before between 1925 and 1927, thereafter motor trains provided a shuttle service between Dorking and Horsham service. Instead of continuing to Leatherhead, they terminated in a new bay created at the country end of platform 3 at the newly named Dorking North. These workings continued until the third rail electrification system was extended southwards from Dorking to Horsham.

Although an instruction was issued that this new third rail must be assumed to be energised on and from 12:05 am on Sunday, 30th January 1938, the full public service did not actually start until May. Included in this electrification was the up siding at Holmwood station, which was also equipped with a 'catwalk' that ran its entire length at platform height. This was to enable train crews to change from one end to the other of a train stabled in the siding. To reduce the voltage drop in the conductor rails that occurs between substations, a track paralleling hut [TP hut] was built at the Ockley end of the up platform, adjacent to the ramp leading down to track level. Other changes included the conversion of the station lighting from gas to electricity, resulting in the replacement of the original LB&SCR cast iron lamp-posts on the platforms with concrete ones in the new SR style, whilst the platforms themselves were raised in height by a few inches and given SR style pre-cast concrete edging.

Thus, when added to the 1926 replacement of the tall wooden LB&SCR signal-posts by SR posts, constructed of re-cycled running rails and steel lattice work, and the removal of the trailing cross-over on the up line beyond the turnpike over-bridge and the associated up advance starting signal, the infrastructure of Holmwood station was starting to take on a distinctly altered and generally more modern overall appearance.

Top: Motor-fitted D1 Class No.2226 propels its train away from the camera. The aspects of the starting and branch signals, together with the tail lamp on the front buffer beam of the locomotive, confirm the direction of travel. The photograph was taken between 1931 and 1940, at an unknown location.

Bottom: E5 class No.B589 at Horsham shed on 5th November 1927. Whilst the similar Class E5x class worked the motor trains between Horsham and Dorking, this engine was also a likely visitor to Holmwood.

87

As ever, throughout this period between the wars, other events at Holmwood station veered from the absurd to the down-right tragic:

Surrey Mirror and County Post
Friday 30th March 1934
LOST AND FOUND

On Friday, March 23rd, at about 4.15 p.m. between Holmwood Station and South Holmwood School, a BROWN BROGUE LADIES' SHOE, makers Randall, brand new. Reward offered. E. G. Rosling, Hillview, Cudworth,

---

Surrey Mirror and County Post
Friday 23rd August 1935
MISS PHYLLIS SHAKESPEAR --
FOUND DEAD IN HOLMWOOD COPSE --
INQUEST STORY OF STARVATION IN SERBIA

Awaiting the result of the analysis of certain organs, the inquest on Miss Phyllis Shakespear who was found dead in Bregnells [sic] Copse, near Holmwood railway station on August 15th, was adjourned by the West Surrey Coroner. The inquest was opened at the Dorking Court House on Saturday. Miss Shakespear was a Civil Servant, aged 43, living at 23a, Bywater-street, Chelsea. She was a daughter of the late Capt. George Courtlant Childe Shakespear of the Indian Army.

Mr. Edward Terrell attended on behalf of Miss Shakespear's family.

The first witness was Mr. George Pelham, a railway ganger, living at 4, Starmount Cottages, Beare Green. He said that, on Thursday, August 15th, while he was examining his length of line he went into a wood near the railway and there he saw a woman lying on the ground. She was lying with one arm out-stretched. A book, called 'The Crimes Club', and a vacuum flask were lying behind her.

The Coroner: Was it anywhere near a track or path? – Pelham: There is a road one side and the railway the other. It is a long strip of copse.

How far from the railway? – About 20 yards from the railway boundary fence. Can you explain how it was not found before? No, sir.

Is it a place that is much used? – Only at blackberry time, when the children go in there.

A Novel, a Flask and a Handbag

P.C. Sturt said that he received information from Dorking Police Station at 8.45 on Thursday morning that a woman's body had been found in a copse near Holmwood railway station. He saw the body in Bregnells [sic] Copse, which is in Capel parish, and which is private land, not common land. The body was in an advanced state of decomposition. The woman was lying on her back on a copy of a London newspaper dated July 13th. Her head was resting on her left arm, and her right arm was over the body. Near by was a novel, a vacuum flask containing a small drop of colourless liquid, and a brown leather handbag.

The Coroner: Did you smell the liquid in the flask? – P.C. Sturt: Yes, there was no smell at all.

In the handbag, he said, there were a cheque book on the Chelsea branch of a bank, a season ticket in the name of Mrs. Phyllis Shakespear and a small tablet bottle containing a colourless liquid.

The Coroner: Was it possible that the body had been lying there for a month or more without being seen? – P.C. Sturt: Yes, sir. Quite possible.

Was see in a natural position? – Yes, just as though she had gone to sleep.

There was nothing to indicate any crime or violence? – No, sir

There was nothing else in the hand-bag, no letter? – No, sir, only some money.

The Coroner examined the season ticket, and said it covered journeys from Charing Cross to Sloane-square, the expiring date being July 22nd.

Mr. Terrell: Was it taken out in the name of Miss Shakespear?

The Coroner: It might be anything. It might even be Mrs. Shakespear. It might quite well be Miss.

Mr. Terrell: I think it was intended to be Miss.

Sister's Search

Miss Barbara Jessie Shakespear, of 23, Fairlawn Court Mansions, Chiswick Park, sister of Miss Phyllis Shakespear said the property found with the body at Holmwood was her sister's property. She had not seen the body owing to its state, and the identification that the woman was her sister was governed by the property found. She last saw her sister on July 10th, when she was very white. Her sister had been suffering from acute headaches and very low blood pressure, which caused great depression. On July 15th, witness found that her sister was not at home at Bywater-street, Chelsea, and witness ascertained that she had been absent from her office since July 13th. The people at the office told witness that Miss Shakespear was away ill. Witness left a note for her sister asking her to ring up. No 'phone message came, and, therefore, witness went again to her sister's home on the Wednesday. The note which witness had left was still there, and the house was still locked up. She then received a message from her sister's office asking if she knew where Miss Shakespear was. The next she heard was from Mr. Waring, an old family friend, who had received a letter from her sister.

Mr Terrell said the letter contained a statement that she was going to take her life.

Miss Barbara Shakespear added that her sister served in Serbia during the war. She took part in the great retreat, and suffered from starvation, which seriously undermined her constitution.

Henry Franks Waring, of 16, Priory-road, W., said the letter from Miss Phyllis Shakespear was received by him on Wednesday evening, July 17th. As soon as he got it, he communicated with Miss Barbara Shakespear, and with the police. Mr. Waring added that he had known Miss Phyllis Shakespear for many years. She had been very depressed and anaemic of late.

Letter Posted From Dorking

The next witness was Dr. Max Davison, Police Surgeon. Before he gave his evidence, the Coroner said: "The doctor is not able definitely to state the cause of death, but he will say there is no violence. The letter is posted from Dorking. It is dated July 13th. This letter is a clear indication of her intention to take her own life by veronal. That is quite clear."

Mr Terrell: She had been working very hard and working overtime and taking home work at night. She had been working very hard in connection with the interior decoration of houses.

Dr. Davison said the body was in such a condition that he was unable to state the cause of death. There were no signs of external violence.

The Coroner: The cause of death I cannot clear up this morning. Certain parts which may help are being analysed. If there is death from violence, such as poisoning, we will have to consider whether anyone has been connected with it, but from the contents of this letter it is clear that she intended to take her own life. I adjourn the inquiry until August 30th.

The inquest was resumed on 30th August and the proceedings lasted less than five minutes. The Coroner said that the report of the Home Office Analyst showed that the organs contained Veronal [the trade name for the first commercially available barbiturate drug] and it appeared that the total amount taken would have been a large and fatal dose - 50 grains or more. A verdict of "Suicide while temporarily of unsound mind" was recorded.

As a footnote to this sad tale, before taking up employment as a clerk in the Ministry of Labour in 1922, Miss Shakespear had served in the British-run Serbian Relief Fund's Front Line Field Hospital.

This had accompanied the Serbian Army's retreat through the Albanian mountains in the winter of 1915-16. It is estimated that some 240,000 Serbs died from the cold, starvation, typhus and enemy action during this withdrawal. For this work, Miss Shakespear was awarded the gold Serbian Cross of Mercy medal.

Another aspect of this story that will carry a resonance with present day residents of Beare Green, bearing in mind that the Surrey Police helicopter is often to be seen flying over Holmwood station, was the use of an autogyro in the search for Miss Shakespear. In the words of an unattributed newspaper clipping of the time:

> An autogyro seen over Holmwood Common, near Dorking, Surrey, yesterday afternoon, is believed to have been a police machine taking part in a search which has been continuing since Sunday, following a woman's expressed interest to commit suicide. The machine hovered over the common for a long time. It was never at a great height, and occasionally when it came down very low an observer could be seen examining the ground through glasses ... ...

Meanwhile, adjacent to Holmwood station itself, the Misses Jackson and Milne had established a small business that not only kept bees, but also supplied teas and honey to walkers and other visitors by train. The following two advertisements set the scene:

Surrey Mirror & County Post
Friday 3rd June 1932

BEEKEEPING. – Bees, Queens, Appliances, Tuition, Hives attended anywhere in Surrey. – Winifred Milne, Holmwood Bee Farm, by Holmwood Station, Beare Green. Telephone Holmwood 63 after 2.30 p.m., except Wednesdays

Surrey Mirror & County Post
Friday 23rd April 1937

BEEHIVE,
Holmwood Station. –
Young Girl or Boy to assist indoors and outdoors.

Class H2 4-4-2 No. 2425 'Trevose Head' lifts its 9-coach London Victoria to Portsmouth, vîa Mitcham Junction, train up the Holmwood bank in fine style.

Designed by D E Marsh, it first emerged from Brighton Works in December 1911 as LBSCR No. 425 and was finally withdrawn by British Railways, as No. 32425, in September 1956. The locomotive carries the post-1926 Maunsell green livery, with white lining, edged in black. Its wheels are also green, with black tyres and centres. The lettering on the tender is in mid-chrome yellow and beneath the word 'Southern' is the post-1931 Central Section engine number. The 660-volt 'third rail' for electric trains is also in place. All of which indicates the photograph was probably taken after 1938 and prior to the time Southern locomotives were painted black in World War 2. The third and fourth vehicles in the rake are Pullman cars.

A contemporary post card showing The Beehive. The advertised 'COUNTRY PRODUCE For the FLAT in TOWN', including eggs and jars of honey, is prominently displayed on trays sitting on the top of the boundary hedges facing the turnpike. The railway station building is behind the photographer.

Chapter 19
THE GOODS YARD

From the outset, it seems quite clear that the LB&SCR had regarded Holmwood as being a station that would be used for the transportation of goods, both into and out of its hinterland. Indeed, the station was sited next to an established brick and tile works and doubtless it was hoped that this enterprise would be the recipient of a steady flow of slack and small coals or other material imported to fuel the kilns, and offer an opportunity to transport the finished product to wherever it was needed, whilst other lucrative common carrier services were developed by the railway company.

Accordingly, a goods yard was designed and built that would satisfactorily serve these purposes until the day it closed. As it turned out, the only changes made during its entire existence were a slight lengthening in the late 19th century of the long siding to accommodate coal staithes and the removal of the good shed in 1958 when it became structurally unsafe.

As can be seen from the 1914 Ordnance Survey plan [overleaf], whilst there was a refuge siding leading from trailing points at the Ockley end of the up platform, all the goods transfer facilities at Holmwood were on the eastern side of the down line. The principal rail access to the goods yard was about 210 yards from the end of the down platform, where a ground frame controlled a set of trailing points into the headshunt. From this, leading back towards the station, a 260 yard long siding curved away to run alongside the boundary fence almost to the approach road leading into the goods yard from the bottom of Station Hill. Approximately half way along this siding was a typical LB&SCR yard crane, with a fixed-elevation heavy wooden jib. Coal staithes were sited between this crane and the buffer stop. The headshunt continued, running parallel to the through lines, and terminated in the cattle dock bay goods platform, whilst a loop ran off from it through the goods shed. To expedite the transfer of horseboxes, cattle trucks or goods vans into and out of the cattle dock and onto or from trains on the down main line, the cattle dock bay was provided with its own direct access to the running line, via trailing points situated beyond the down starting signal and just short of the trailing cross-over between the down and the up running lines.

However, whilst the coming of the railway allowed brick and tile making concerns to flourish at Ockley and Warnham, none of the OS plans published after 1896 show any trace of the brick and tile works adjacent to Holmwood station, other than the flooded pit that had formerly used for clay extraction.

Fortunately, many other businesses in the neighbourhood quickly recognised the benefits of the railway for the general transport of goods, and perhaps none more so than local farmers.

To show the sheer variety of agricultural merchandise that passed through the Holmwood station goods yard in connection with just one family dairy farm, a digest [see overleaf] has been created from relevant entries contained in the New House Farm bailiff's daily diary [1908-1924], the farm ledger [1903-1918] and the wages book [1917-1926].

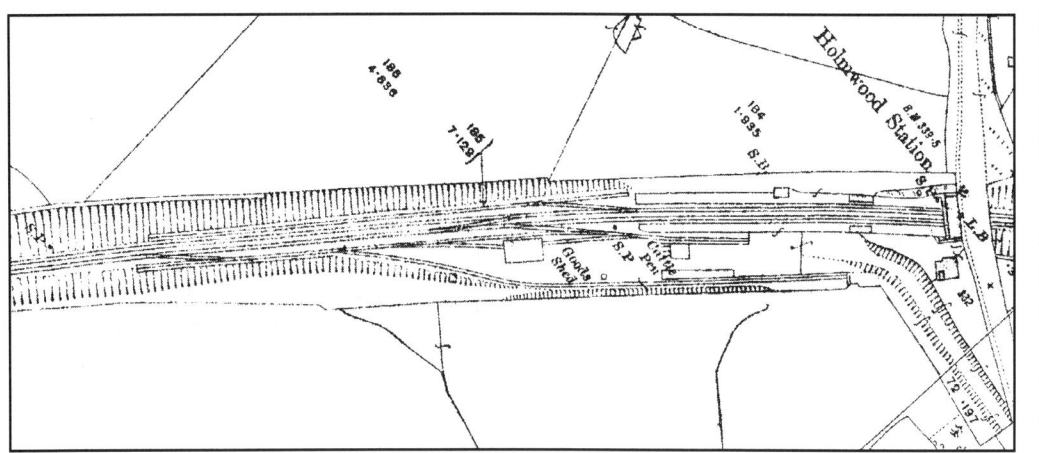

> **New House Farm Wages Book,
> Friday 6th March 1926**
> Tip to Porter for unloading soot: 1 shilling.

Continuing this farming theme, a local agricultural auctioneer, Mr Frederick Crow, had started to conduct sales of live and dead-stock in the vicinity of Holmwood station from about 1890, if not earlier:

Over the next 45 years, including the period 1914-18, these agricultural sales grew increasingly more popular with local farmers and profitable for Mr Crow. The

---

Goods imported by New House Farm, Newdigate collected from railway wagons at Holmwood Station:

Fertilisers - Basic slag [At the time, considered by many to be the best grassland fertiliser. Also useful as a 'liming' agent on the Weald clay]; Super-phosphate [Another popular phosphatic manure of the time, that induced good root development; Kainite [A source of potassium, vital for the development of flowers and fruits]; Limphos [The brand name of a superphosphate fertilizer]. These came in the railway wagon-load.

Cattle - Including calves; heifers, in-calf cows, milking cows and bulls

Horses - Including colts, cobs, cart horses and cart mares, ponies, vanners and mules. [The boxes for these animals appear to have been attached to passenger trains.]

Stock feed - Oats; dairy meal; cotton seed cake; linseed oil cake; middlings [A grade of meal]; coarse middlings; Molassine cut meal; Albion dairy cake; Albion dairy meal; gluten meal; Quaker oats; Quaker oat meal; oats; toppings; mill sweepings; soya bean cake; Thorley's cake; Thorley's dairy meal; Egyptian cotton cake; linseed; niger seed; Calthorp's dairy cake; barley meal; cotton cake; biscuit meal; Bombay cotton cake; treacle; maize; Shirleys dairy meal; Shirleys prize cake; cocoa nut cake; palm nut cake, palm nut meal; bran; broad bran; horse corn & poultry corn; mixed corn; wheat screenings; pig meal; swedes; ground oats; beans; split beans; chaff; cracked maize; 'Golden Tankard', 'Large Globe' & 'Red Intermediate' mangolds; mangold wurzels decorticated meal; spent brewer's grains, including ale grains, porter grains and distiller's grains.

Manures - Straw manure; long dung; straw & shavings manure; sludge manure; [Although the farm produced enormous quantities of manure from the cattle yards, it seems that it could always use more, again by the railway wagon-load.]

Sundry Items - ashes, coarse ashes, clinkers, brick rubble; sawdust; wood chips; wood shavings; empty milk churns; thatching straw; wire hurdles; flints; wooden shed; hay press; turkeys [in a crate]; a goose; cattle cribs; sheet iron; chestnut fencing; tarpaulin covers; bricks; fencing rolls; timber gates; corn & seed drill; new tank for water cart; wheat straw; oat straw; spring oats [seed]; winter oats [seed]; wheat [seed]; water tank; hay cord; cleft chestnut fencing; new poultry houses; best Wallsend house coal [from Tyneside, for domestic purposes]; Linby hard steam coal [from Nottinghamshire] and other steam coal [for steam-ploughing and threshing];

Goods exported from New House Farm, Newdigate, delivered to Holmwood Station:

Milk - This was the principle product, being sent in churns [twice a day in summer, daily in winter] to an associated dairy in Peckham.

Cattle - Including calves; heifers, in-calf cows, milking cows and bulls

Horses - Including colts, cobs, cart horses and cart mares, ponies, vanners and mules. [Again, the boxes for these animals appear to have been attached to passenger trains.]

Hay - Initially this was despatched in trusses, with 63 trusses weighing 1 ton. Later, the farm at first borrowed and then bought a hay press, which allowed the hay to be made into bales for easier transport; 40 bales of pressed hay weighed 1 ton. Hay is bulky for its weight: a horse-drawn waggon could only carry one ton of hay, and two tons filled a railway truck.

Sundry Items - Willow; firewood [bagged and loose]; pea & bean sticks; wooden poles; potatoes; returned sacks.

convenience and proximity of the Holmwood station goods yard was a major factor in the growth of this business. Whilst sales of individual properties or businesses were conducted on an ad hoc basis, the stock sales settled into a regular half-yearly pattern at Garston's Barn. However, after 15 years, the 31st sale was moved in September 1924 to a new sales ground nearer to the station. This had permanent sorting and marking pens, which was a great convenience for the lotting of livestock on entry and for expediting delivery to their new owners.

The account [see right] of one of Mr Crow's sales is typical and the number of railway wagons that had left the goods yard by the end of the day is worthy of note, as are their destinations.

**The Sussex Express,
Surrey Standard,
Weald of Kent Mail,
Hants and County Advertiser,
Saturday 13th September 1890**

BEARE GREEN, HOLMWOOD,
close to Holmwood Station, L.B. & S.C.R.
10 EXCELLENT DAIRY COWS and HEIFERS (in calf or in profit), three weanyear heifers, two sows with pigs, sow in pig, about 80 head of poultry, nearly new winnowing machine, and three stacks of capital Meadow and Seed HAY, estimated at 25 loads,
*FOR SALE BY AUCTION, BY
MR. FREDK. L. CROW,*
Under instructions from Mr. John Fowler,
on the premises, at the Duke's Head Inn, on MONDAY, SEPTEMBER the 15th, 1890 at Two o'clock precisely.

May be viewed the morning of sale, and catalogues may now be obtained at the Inns and Hotels in the neighbourhood or of
Mr. FREDK. L. CROW,
Auctioneer and Valuer, South Street, Dorking.

**Dorking & Leatherhead Advertiser,
Saturday 6th May 1916**
HOLMWOOD STOCK SALE

The sum of £48 15s. was the handsome figure paid by Mr. W. Barker for the best fat bullock at the half-yearly sale at Holmwood Station, conducted by Messrs. Crow on Monday, a primes half-bred Sussex steer from Ewood farm, Holmwood. There were eight Devon and other beast [sic], which made £40 and over. Seventy-four fat beast [sic] were sold, and a grand lot they were, the majority being well-fed half-bred Sussex, a cross which is deservedly attaining a great reputation in the district. One hundred and fifty head of store cattle passed through the ring, evoking keen competition and realising up to £26 10s. apiece; generally speaking they were a very nice run of bullocks, but some were a little bare, doubtless having suffered from the long winter and shortage of hay. The 40 head of dairy cattle found ready buyers, the top price of £32 being secured by Mr. Winfield for his half-bred cow due with her third calf. Several useful bulls completed the tale of horned stock. Fat tegs were a good trade at 56s. to 82s., and Mr. Teasdale's 60 half-breds averaged 72s. apiece. One hundred and twelve pigs were sold, mostly at long figures, hogs reaching nearly £10, porkers 75s. to 90s., and store pigs 20s. to 32s 6d. Twenty-six horses were offered, and nearly all changed hands at satisfactory figures.

As usual, a quantity of dead stock and a number of poultry were catalogued, and in consequence of the number of lots the sale commenced punctually at 10.30 o'clock, the last lot being knocked down at 5 o'clock. The delivery of the stock was being carried out throughout that time, and nearly 30 truck loads left Holmwood station that day, whilst a few more were loaded there and at Dorking the day following, being consigned to all parts of Surrey and Sussex, and as far distant as Bucks and Northampton.

However the transport arrangements did not go quite as smoothly after the Autumn Sale in the previous year [see overleaf]. In passing, at this particular sale, a run of draft Southdown ewes from Mr Cuthbert Heath's farm at Anstie Grange made from 41s. to 53s. apiece.

The success of the railway as a common carrier was also reflected in the various enterprises set up by local individuals to transport coal and other goods from the railway station to more outlying areas. However some

> **Dorking & Leatherhead Advertiser,
> Saturday 11th September 1915**
> *HOLMWOOD STOCK SALE*
>
> What has become almost proverbial "Holmwood sale weather" was enjoyed on Monday, when the half-yearly fixture was held, and there was a particularly large number of both entries and purchasers, between £4,000 and £5,000 worth of stock changing hands …
>
> … The sale, which comprised some 230 lots was concluded before six o'clock. Unfortunately, military requirements prevented trucks from being at Holmwood Station on the Monday, but the auctioneers, Messrs. Crow, undertook to keep the stock as long as was necessary, and nearly all were put out on rail at Holmwood or Dorking (S.E. Railway) the following day.

of these ventures were perhaps more profitable than others and, again, various contemporary newspaper advertisements provide an insight into the size and nature of these businesses:

> **Surrey Mirror, Friday 17th June 1904**
> *SALES BY AUCTION*
> *MESSRS. WHITE AND SONS.*
> *ON MONDAY NEXT.*
> ------
> *HOLMWOOD STATION, SURREY.*
>
> Messrs. White and Sons are favoured with instructions from Mr. J.W.Turner to sell by Auction, on MONDAY JUNE 20th, 1904, at 2.30 o'clock, the *STOCK-IN-TRADE OF A COAL AND COKE MERCHANT, viz:-*
> 3 VAN HORSES, CHESTNUT PONY, 3 good SPRING VANS (two with hoops and tilt) LIGHT PONY SPRING VAN, VAN AND PONY HARNESS, QUANTITY OF COKE, FERN AND HOOP CHIPS; 4 SETS OF COAL SCALES AND WEIGHTS, 175 COAL SACKS:
> Ladder, Corn Bins, Stable Tools, Shovels, Wheelbarrow, Sieves, Pails, etc.
>
> Catalogues may shortly be obtained at
> the White Hart Inn, Holmwood Station, and of Messrs. White and Sons, Auctioneers and Valuers, Dorking and Leatherhead.

> **Surrey Mirror
> Friday 13th December 1929**
> *SALES BY AUCTION*
> *MESSRS. CROW.*
> *SHORT NOTICE OF SALE.*
> By Direction of the Executors of Mr. W.F.J. Brown, decd.
> HOLMWOOD STATION, SURREY.
> About 4 miles South of Dorking.
> Messrs. Crow are favoured with instructions to Sell by Auction, in Lots, at THE RED LION HOTEL DORKING, on MONDAY, DECEMBER 23rd, 1929, at 4 o'clock, THE FREEHOLD PROPERTIES comprising:-
> "STARMOUNT VILLA",
> The COTTAGE adjoining, the GARAGES, STABLING, AND YARD
> Together with the
> GOODWILL OF THE COAL MERCHANT'S & CARTING CONTRACTOR'S BUSINESS.
>
> Particulars and Conditions of Sale can be obtained of the Solicitors, Messrs. Hart, Scales, and Hodges, Dorking and Leatherhead, and of the Auctioneers, 76 South Street, Dorking. Phone 176.

A variety of coal and coke merchants operated from the coat staithes in the Holmwood goods yard at any one time, including a Mr. Jackson who supplied Newdigate village in the years before the Great War and the business run the Hoad family in South Holmwood. This latter concern was started by John Hoad, who had previously been the first Porter at Holmwood recorded in the LB&SCR Staff Register. Evidently he spotted a likely business opportunity, as he resigned from railway service in 1868 and started a horse-drawn transport undertaking to carry passengers and their luggage to and from the station. Although John Hoad died in 1923, his carrier business, which had included transporting drums of petrol delivered to Holmwood station yard for Mr Pethick Lawrence's motor cars, was continued by his son, Arthur Hoad. He used motorised transport to carry a wide variety of railway-born goods, including coal, bricks, and products from the Schermuly pistol rocket factory in Newdigate.

Printed Ephemera - 3

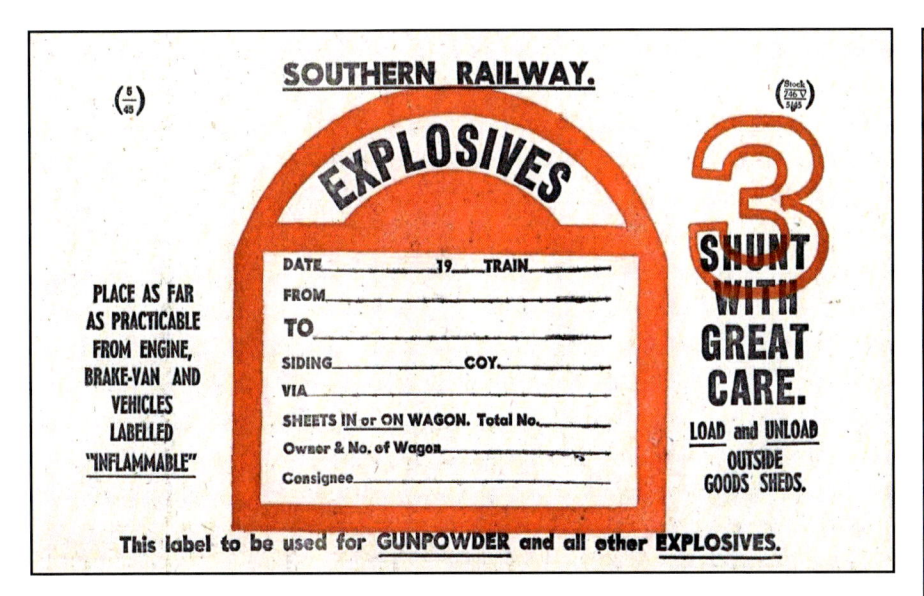

Southern Railway and British Railways wagon labels relating to goods traffic at Holmwood.

95

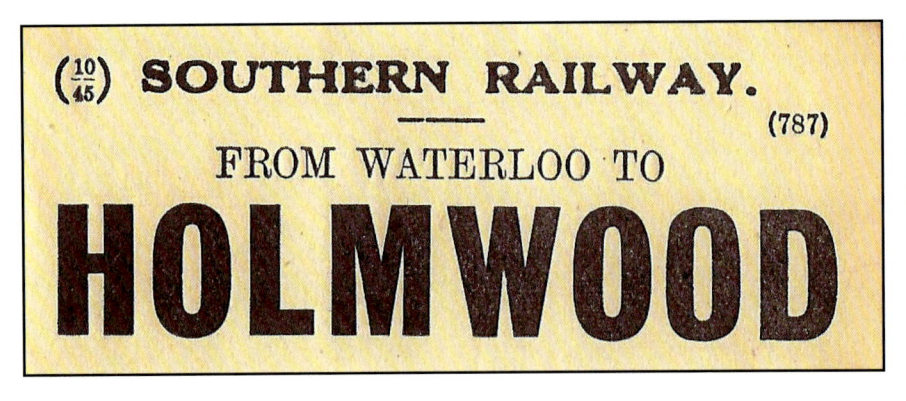

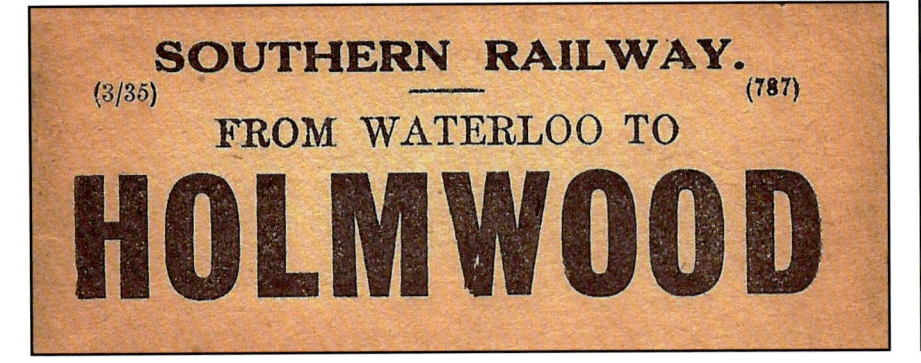

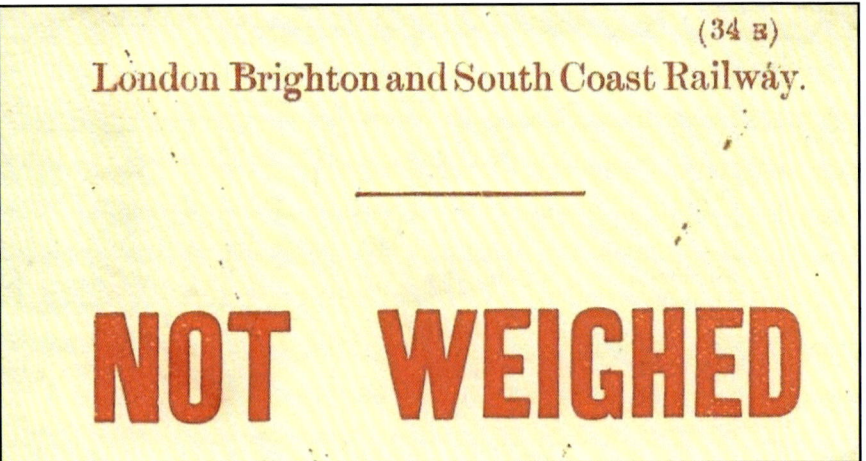

Printed Ephemera – 4
Miscellaneous LB&SCR and Southern Railway goods and luggage labels.

To accommodate their growing business and to have a conveniently located test firing range, the Schermuly firm had moved from Cheam to a larger site at Parkgate, on the outskirts of Newdigate, in 1937. At the time, their principle product was a pistol from which a rocket propelled line could be fired from ship to shore. Bearing in mind that the apparatus would be used in extreme conditions, it needed to have a long shelf life and be small; light; easily aimed; accurate; waterproof; safe and simple to use - so simple in fact that it could be, as the firm demonstrated in its advertising, "fired by a child".

Whilst the propellant used in the Schermuly rockets was delivered to Holmwood station goods in purpose-built, steel, railway gunpowder wagons, for reasons that are not entirely clear, their finished products were delivered to customers by the Southern Railway in ordinary, wooden, covered goods vans.

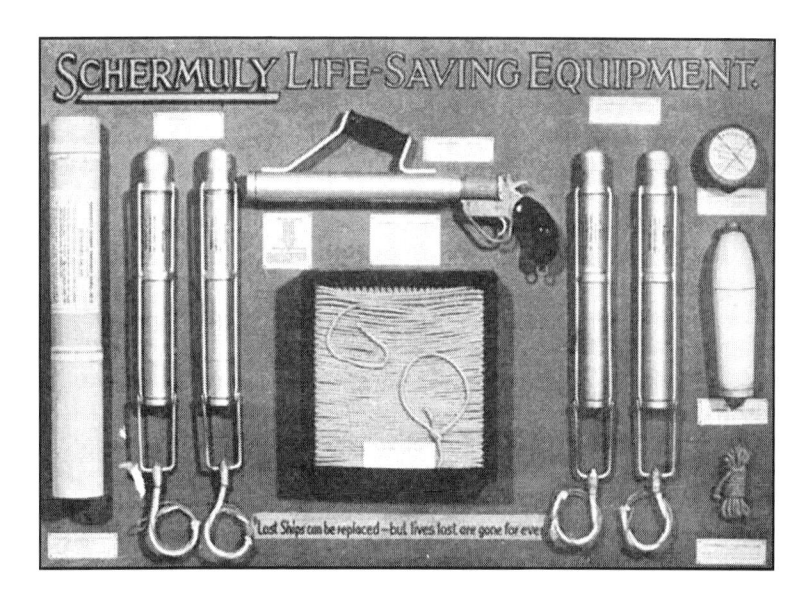

Left: Arthur Hoad, 'cartage agent for the Southern Railway', and his lorry.

Right: The Schermuly Pistol Rocket Apparatus in a display case bearing the slogan, "Lost ships can be replaced – but lives lost are gone forever".

Chapter 20
INTO ANOTHER WORLD WAR

Since late Victorian times, the basic passenger train service from Holmwood had been two trains per hour, in each direction. However, following the extension of the third rail electrification scheme from Dorking to Horsham in 1938, the Southern Railway took the opportunity to extend the hourly [6.48am to 9.48pm] Waterloo to Dorking service to Holmwood. These additional trains terminated at the down platform and the empty stock was then shunted across to up line before being reversed into the up refuge siding, thereby clearing the line for other traffic until it was time for it to form the next up service back to London.

This excellent train service did not pass unnoticed, either by those new commuters who wished to take advantage of the countryside surrounding Holmwood station, or by those who were willing and able to sell building plots to them.

Throughout the 1930s, ribbon development had started to take place along the turnpike in both directions away from the station. In the latter years of that decade a rather more ambitious, albeit piecemeal, development scheme was started by a local landowner on what was then known as the White Hart Estate. On this site, individual plots were sold off to individuals who wished to construct their own dwellings to suit their particular fancy.

In January 1938, the Surrey Mirror newspaper reported that the Surrey County Council had approved the making-up of the various access roads within the White Hart Estate under the Private Street Works Act 1892.

As the prospect of war loomed ever larger in 1939, perhaps the most obvious change that took place at Holmwood station was the imposition of 'the blackout'. This was ordered by the Minister for Transport on 1st September, two days before the actual declaration of war itself. As part of this requirement, railway carriages had their windows screened with blinds or painted black and blue light bulbs were fitted. For safety, the edges of railway platforms were painted white. Throughout the war, the blackout was strictly enforced and signalmen were instructed to stop trains that were showing a light. Instructions were also issued to the travelling public.

The 'phony war', a period of relative anti-climax that lasted until the early summer of 1940, was brought to an end by the evacuation of the British Expeditionary Force from Dunkirk and the start of the Battle of Britain. As the German bombing offensive intensified over London, those wishing to escape the Blitz presented an unexpected business opportunity to at least one local land speculator [see advertisement on page 101].

DURING BLACKOUT HOURS

KEEP ALL BLINDS DRAWN

KEEP ALL WINDOWS SHUT except when it is necessary to lower them so that passengers may open doors to alight.

MAKE CERTAIN that the train is at the platform and that you alight o the platform side.

WHEN LEAVING THE CARRIAGE close windows, lower blinds again, and close the door quickly. Doors leading to corridors should be closed at once after passengers have left the compartment and blinds readjusted, unless outside corridor windows are completely blacked out.

DO NOT TOUCH ELECTRIC LIGHTS except those under passengers' control. Pleases witch these off when not in use.

ALL NECESSARY LIGHTS will be switched off in the event of an Air Raid warning.

These arrangements are experimental and their continuation depends on
YOUR CO-OPERATION.

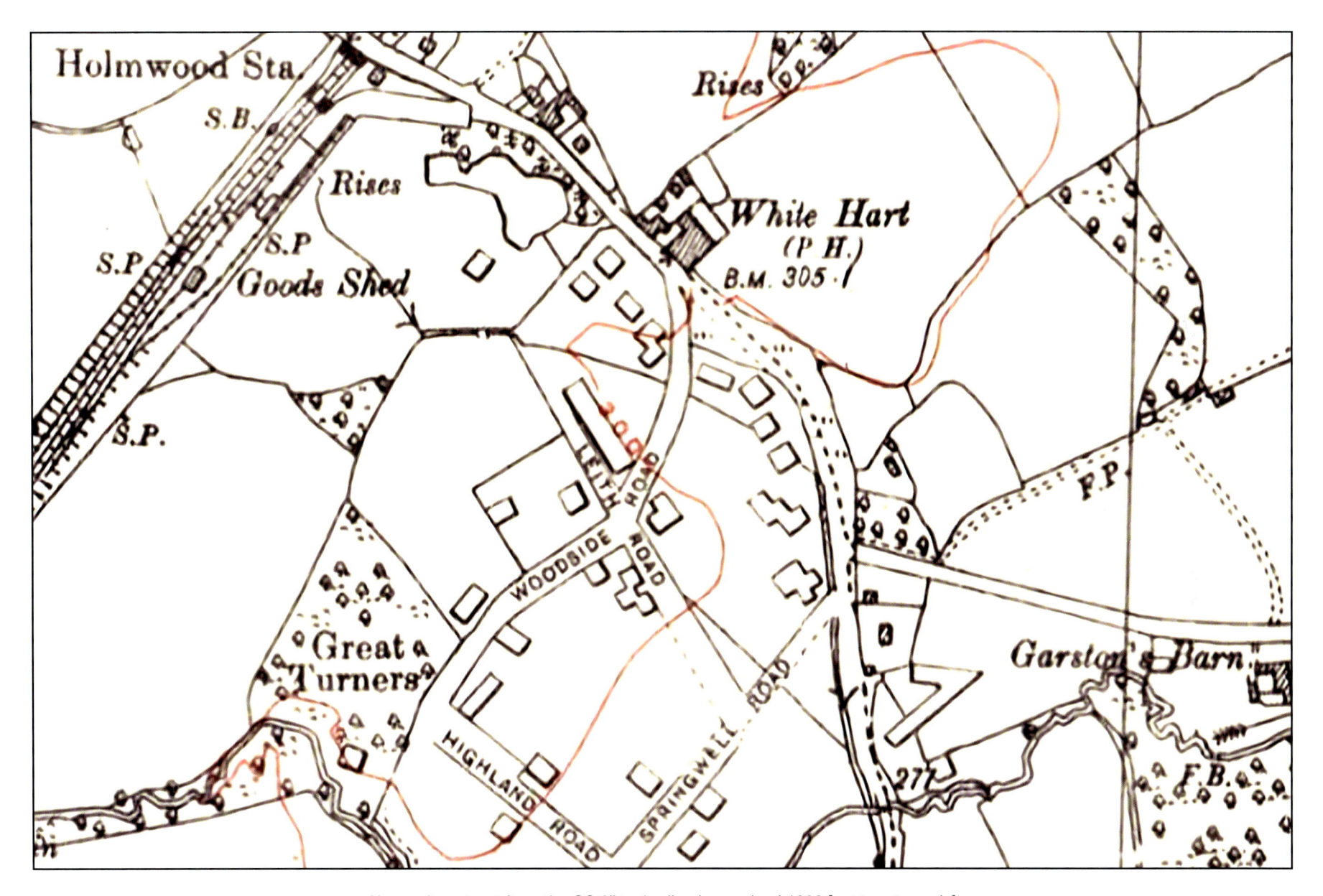

Above: An extract from the OS 6" to 1 mile plan, revised 1938 [not true to scale].

The embryonic road layout of the White Hart Estate is clearly shown, with access roads leading directly off the Dorking to Horsham turnpike.

Above: An advisory poster from the Second World War. It may seem to be stating the obvious, but the blackout made conditions very hazardous for railway travellers previously used to peace-time levels of lighting.

Top: An artist's impression of a German Dornier bomber being brought down by a Schermuly Parachute and Cable [P.A.C.] rocket, fired from a ship, during World War 2.

Bottom: Just as in the Great War, railway companies issued lapel badges to their employees to show that they were contributing to the war effort. This Southern Railway example is of the type worn by railwaymen at Holmwood. Irrespective of the issuing company, all these badges carried a silhouette of a Great Western Railway 4-6-0 locomotive. This badge was made in Birmingham by Fattorini & Company Ltd.

> **Surrey Mirror and County Post**
> **Friday 10th May 1940**
> *HOUSES AND LAND FOR SALE*
> Land, 40ft. by 250ft., more available, Freehold, unrestricted; in shadow of Leith Hill; five minutes from Station; roads made; all services laid; £120. – Maybelle, White Hart Estate, Beare Green.

As the war progressed, the superb train service to and from London at Holmwood station ensured that 'homesteading' on the White Hart Estate continued as more folk sought to 'live in Surrey, free from worry', but there are hints that standards there may have started to slip:

> **Surrey Mirror and County Post**
> **Friday 14th November 1941**
> GOAT. – Well-bred HORNLESS BILLY for Sale, 18 months; good stockgetter; 3 guineas. – Dewdney, Hill View, White Hart Estate, Beare Green.

Furthermore, concerns were being expressed by the Dorking and Horley Rural District Council about the water supply to estate and the inevitable consequences for public health:

> **Surrey Mirror and County Post**
> **Friday 6th February 1942**
> *Water Supply at White Hart Estate*
> Concerning a water supply to the White Hart Estate at Beare Green, the Clerk reported that replies received from the owners were unsatisfactory; frontagers owning 770 feet only out of 3,784 feet were in favour of the main extension and paying not exceeding 2s. and 6d. a foot. The Committee were asked to reconsider the subject.

The generally slipshod way this estate was developed during the war years is a theme that will be expanded upon further in a subsequent Chapter.

Whist Holmwood station does not seem to have been a direct target for the enemy, the railway staff would all have been acutely aware of the fact that all sorts of dangerous objects, such a jettisoned bomb-loads and shot down aircraft, were likely to fall out the sky at any moment without warning – as they did in the general vicinity throughout the War.

As the War progressed, the Schermuly factory diversified its production from solely life-saving equipment towards more military and ultimately aggressive fields of endeavour. As in peace-time, the gunpowder and other explosive materials used as propellants came into Holmwood by train for onward delivery to Parkgate and the finished products were delivered in the same manner. The result was that the sidings in the goods yard were usually packed to capacity throughout the War.

Innovative Schermuly products during this period included 7 million candle power Target Identification Flares, used by pathfinder squadrons to mark targets for bombing raids; Parachute and Cable [P.A.C] rockets, used to launch a hawser designed to foul the wings of attacking aircraft and Grapnel Rockets used by troops assaulting steep cliffs on Omaha Beach on D-Day, 6th June 1944. Yet perhaps their most technically impressive invention was the Air Sea Rescue Discharger, an inflatable life-raft that, when dropped among ditched crews swimming in the water, opened automatically, inflated and then fired rocket propelled lines into the water, thereby allowing the men to haul themselves towards and then into the life-raft.

Prompted by the allied landings in Normandy, the first V-1 flying bomb [a so-called 'Vengeance Weapon', Vergeltungswaffen] was launched at London on 13th June 1944: a campaign that continued until October 1944. Again, whilst at least one V-1 did land in the vicinity, Holmwood station remained unscathed. However, as recalled by former Dorking resident, Brian Buss, the up platform offered a grandstand view of these missiles flying towards London on one evening as he was returning home with his girl-friend. They had been to a dance at Beare Green Church Central School [now The Weald Church of England Primary School], which had opened in 1940.

By any reckoning, the potential for catastrophe and a major conflagration in the goods yard at Holmwood station had been very great indeed.

Chapter 21

STATE INTERVENTION

By the end of the Second World War in 1945, the railways of Britain were run down and desperately in need in need of investment. Alas, much the same might also have been said for parts of Beare Green, as the White Hart Estate found itself in the news yet again:

> **Surrey Mirror and County Post**
> **Friday 9th August 1946**
> *Beare Green*
> UNSATISFACTORY HABITATIONS - Complaints were made at the last meeting of the Dorking and Horley Rural Council about two habitations on the White Hart estate, Beare Green. One was a converted motor-car. The other was a dwelling alleged to be in an advanced state of dilapidation. The council decided to take steps to get the motor coach removed and the other building improved. The Council were told that there was on the White Hart estate a wooden shed divided into two rooms, the larger being 11 ft. 6ins. by 9ft. 3ins., occupied by a man and wife. Means of cooking were unsatisfactory. Water supply was from a standpipe 120 ft. away. Sanitary accommodation was primitive. The structure was in fair condition. The couple were paying £1 per week in rent. No rates were paid as up to now the shed had not been assessed. The woman did not want to appeal about the rent in case notice to quit was given. Neither did she wish to make any representations about the condition of the shed.

The various problems caused by the ramshackle nature of some properties on the White Hart estate, essentially a direct product of the excellent train service that allowed Londoners of virtually any social class to escape the Blitz, went on to influence local government

> **Surrey Mirror and County Post**
> **Friday 4th November 1949**
> *RURAL DRAINAGE*
> *DORKING AND HORLEY R. D. C. FACE BIG SCHEMES*
> … The Council directed the Surveyor to supply details of sewerage schemes for Smallfield-road, Horley; the village of Buckland; the White Hart estate, Beare Green; and Misbrooks Green, Capel. When submitting these schemes to the Ministry of Health the Minister is to be asked to take their cost into account when further consideration is given to the question of a contribution towards the cost of water supply schemes …

thinking about Beare Green for decades afterwards. Moreover, that same quality of train service had also been noticed by Central Government when considering the post-War reconstruction of the London conurbation. In the 'Greater London Plan 1944', produced by Professor Patrick Abercrombie for the Minister of Town & Country Planning, Holmwood was identified as a site for a suitable 'satellite' town, adjacent to the proposed new airport at Gatwick. It stated:

> **HOLMWOOD (SOUTH OF DORKING)**
> This site should be residentially attractive, possessing good views especially towards the North Downs. It is served by the London-Horsham main line. Flat land is available adjoining the railway for industry. There is a small amount of scattered inexpensive development. Accessibility will be greatly improved by the proposed Brighton express arterial road. A possible objection to this site is its proximity to Gatwick Aerodrome which may be very much expanded. As regards sewerage, the aim should be to keep clear of the valley to the north, and use only the valley to the south of the town … …

Apart from the governmental fixation with sewerage, bearing in mind the description of Holmwood as a town and the comments about flat land available for industry or the road route to Brighton, one does wonder if the planners drafting this Report had actually visited the site. Indeed, even more surprising, bearing in mind that Holmwood is in the Green Belt, was the fact that Professor Abercrombie had been closely involved in the founding of the Council for the Preservation of Rural England in 1926 and had once served as its Honorary Secretary. Thankfully, all these alleged attributes were found to be actually available in what became the 'New Town' of Crawley.

However, the national railway system as a whole was in a parlous state. Hence, as part of its policy to nationalise public services, the post-War Labour Government, led by Clement Attlee, took the railways into public ownership. Thus, under the provisions of the Transport Act 1947, British Railways came into being on 1st January 1948.

Unfortunately, the realities of a 'Brave New World' public transport utopia did not always meet the expectations of the individual and the letter [right] neatly sums up the hopes and shattered dreams of many a commuter in the decades that have followed since it was first written.

Nationalisation did not have much immediate effect on the nature of the rolling stock used for passenger services, as the electric multiple unit trains used on the line had only been put into service in 1941. They were purpose-built, 4-car suburban sets [hence their classification name, 4-Sub] designed by Oliver Bulleid, the Chief Mechanical Engineer of the Southern Railway. With their distinctive curved sides and ability to seat six third class passengers across their width, 4-Sub units continued to be produced by the Southern Region of British Railways until 1951.

On Thursday, 6th November 1947, the 4-45 pm up passenger train from Holmwood to Waterloo, formed of a 4-SUB unit, collided with another train in darkness and exceptionally thick fog near Motspur Park at about 5-50 pm. Four passengers were killed in the accident. The driver of the train from Holmwood, Motorman A T Drawbridge, was detained in hospital with severe bruises and shock. He was fortunate to escape with his life, as his cab was completely wrecked. Eight of the passengers were also taken to hospital. At the subsequent enquiry neither of the drivers was found to be at fault and the cause of the accident was attributed to a disregard of fog-signalling instructions and the Rule Book by a railway ganger.

> **Dorking & Leatherhead Advertiser**
> **Friday 27th January 1950**
> *OUR READERS' VIEWS – A TRAVELLER'S COMPLAINTS*
> Sir, - British Railways run their 5.27 and 6.27 evening trains from London so as to arrive four to seven minutes late at Holmwood station, thus neatly missing the Capel bus, which is supposed to, but often does not, run in connection. On a Saturday morning at Dorking bus station I was refused admission to a 414 bus to Horsham, although nobody in the bus was standing. Yet it is common knowledge that several passengers invariably alight at the next stop to the southward. Complaints to the bus authorities bring a temporary improvement which soon disappears; complaints to British Railways, by all appearances, receive no attention at all. Is it too much to expect that the local public transport services should imbue themselves with a better spirit of public service?
>
> Yours etc., H. G. Evans, Capel End, Capel

From 1960 onwards, the 4-SUB units on Holmwood services were replaced by four-car sets with electro-pneumatic brakes [hence their classification 4-EPB] and these lasted well into the 1990s. Both these classes had 'slam-doors' that could be opened by passengers whilst the train was moving, whilst their door windows were originally supported by a leather strop, later replaced by a metal friction-bar, which allowed it to be dropped, when open, into a recess in the door panel beneath.

Since Victorian times, railways had been deemed to be 'common carriers'. This obliged railway companies to carry any cargo offered to it at a nationally agreed charge. The original intention had been to stop railway companies "cherry picking" the most profitable freight business whilst refusing to carry less profitable consignments. This had been a very necessary measure when railways effectively had a monopoly over land transport, but with increasing competition from road hauliers, who could be selective, it put the railways at a distinct disadvantage. This common carrier requirement was abolished by the Transport Act 1962, which effectively meant that even less goods traffic was carried by rail.

There had been another blow to goods traffic at Holmwood in 1958. The original goods shed, dating from the opening of the line in 1867, was found to be structurally unsafe and it had to be demolished. To reflect the now diminished goods traffic, a much smaller corrugated iron structure and a short unloading dock was built to replace it. Nevertheless, a goods train ran from Horsham every day, except Sundays, until 1959, when the service was reduced to Mondays, Wednesday and Fridays only. These trains were hauled by steam locomotives until 1963 and thereafter by an 0-6-0 diesel shunter until the goods yard was finally closed in 1964. During the early summer of 1965, the ground frame controlling the points into the goods yard, the crane and replacement goods shed were removed, and the track in the goods yard lifted. However, the cattle dock spur and its connection to the down line were retained for use by the engineer's department.

The passenger service was also in decline. After the summer of 1976, the up siding was no longer used for berthing trains. In November 1977 this siding was also lifted as passenger trains no longer terminated at Holmwood at all.

In 1968 an overhaul and redecoration of the station took place. One unexpected outcome of this work was the discovery of graffiti in one of the station shelters from 67 years earlier, placed there by LB&SCR painters and carpenters on their last day of railway employment before going to fight the Boers in South Africa in 1901. Less satisfactory was the removal of the roofing and upper side-cladding to the footbridge and steps. Ostensibly done to reduce costs, the opposite was the case and rot got into the woodwork, leading to the eventual closure of the station building. After being shut for six years, the 119-year old building was demolished during the first two weeks of April 1986. Its rubble was taken to Bregsells Farm to be used as hard core in the farm yard and in various muddy gateways. What an ignominious fate for a building that had served the community so well for so many years.

Government control of the railways finally came to an end with the privatisation of the network in 1997. The steady decline of Holmwood station during its ownership by British Railways is graphically illustrated in the photographs that comprise the remainder of this Chapter.

Opposite: Despite cheeky faces peering from the footbridge and a carriage window, on Sunday, 7th June 1959, Holmwood station is giving a very fair impression of 'Adlestrop' as described in the 1917 poem by Edward Thomas:

*"No one left and no one came
On the bare platform."*

The train comprises 4-Sub unit, No. 4127, built in 1946, and is waiting to depart from the up platform for Dorking and London. The footbridge still retains its overall cover and sliding sash windows.
Also worthy of note are the red-shaded oil tail-lamp hung on a lamp bracket at the rear of the train and the wooden privy outside the signal box.
The 4-Sub units could always be distinguished from the later 4-EPB units by the draw-hook and screw coupling in the centre of the buffer beam. The 4-EPBs had buck-eye couplings.

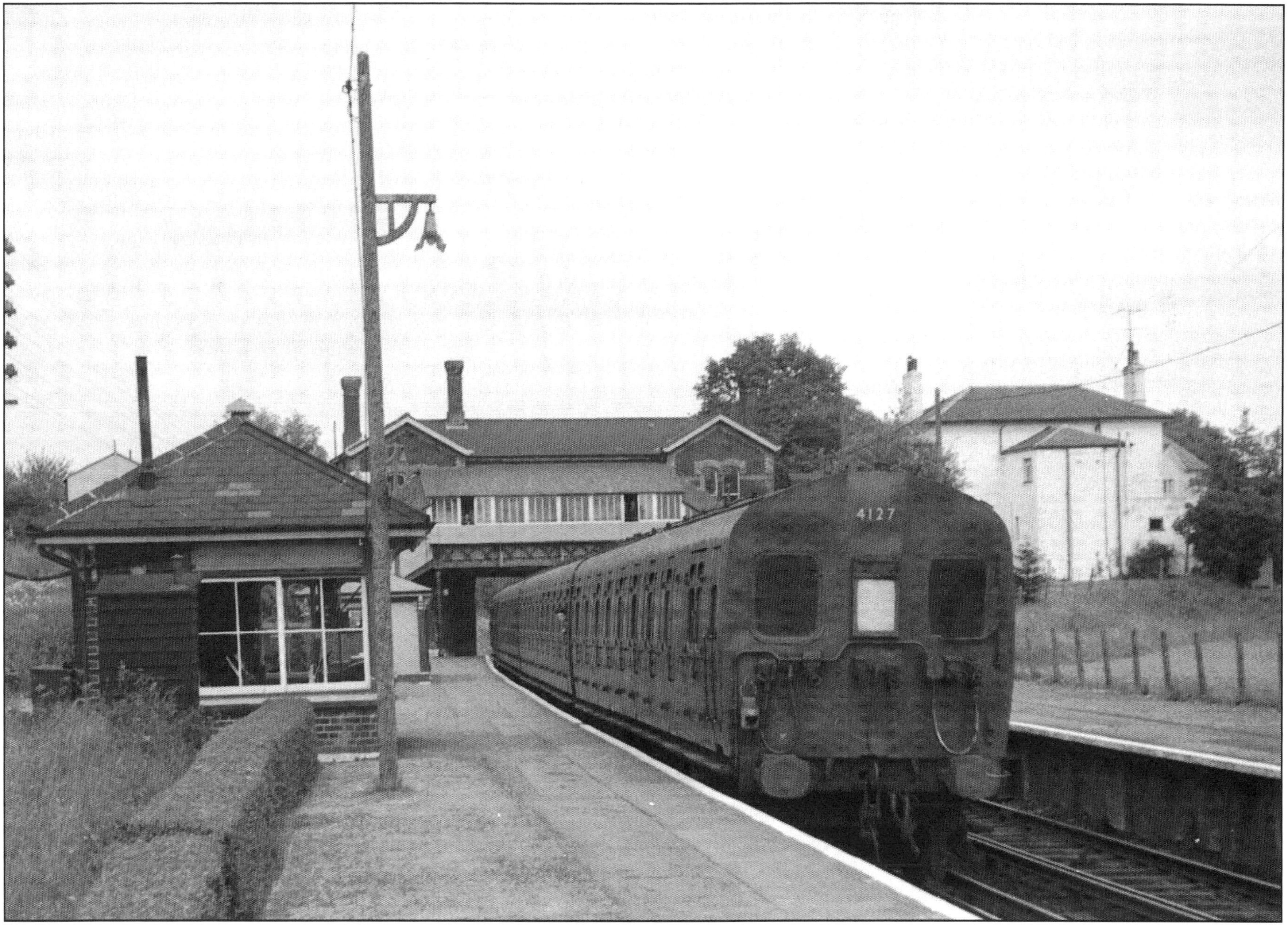
4127

This pair of photographs was taken at Holmwood sometime between 1958 and 1964 and almost certainly on the same day.

Left: Taken looking in the up direction, this is a particularly clear view of the rear of the station building, the covered footbridge and the steps leading down to the platform. At the foot of these steps was a wicket gate, placed there to ease the collection or checking of tickets. This part of the station had not materially altered since the days of the LB&SCR.

The door to the lamp room is particularly prominent to the left of the fire buckets hanging on the wall, with the door to a disused WC to their right. The glass sides of the covered way at road level have two sliding casements.

Top: With the camera facing the other way, in the down direction, it is this view that provides the evidence to date these photographs.

The original 1867 goods shed was a substantial building, but it was found to be structurally unsafe in 1958 and immediately demolished. Whilst the goods shed is not there, there are goods wagons in the yard, which did not close to traffic until 1964. Thus there is a six year period during which these images could have been taken.

The seat, which looks a lot more comfortable than its modern equivalent, and the large station sign are common reference points in the two photographs.

The original LB&SCR shelter on the down platform was demolished in the late 1980s and replaced with a 'bus shelter' type structure.

This photograph, probably also taken during the period 1958-64, shows Holmwood station looking particularly neat and tidy. The privet hedges are well-trimmed and the grass is cut. Points of interest in this view are the foot-crossing for railway staff; the point-rodding and signal wires emerging from the 'leadaway' that gives access to the locking-machinery room beneath the signal-box; the three varieties of station name boards and the number of individual telephone lines carried by the telegraph pole. The rural nature of the hinterland lying to the West side of the railway line is very apparent.

A moment of anticipation at Holmwood on 28th March 1965. The down starting signal is off and the goods yard loading gauge frames the Station Master's house perfectly. But already debris is starting to accumulate in the goods yard, which closed in 1964 and the livestock pens have already been removed from the cattle dock. The rectangular bulk of the track paralleling hut [T.P.hut] is particularly prominent at the country end of the up platform.

BR Standard Class 5 4-6-0, No.73022, hurries through Holmwood with a Railway Correspondence and Travel Society rail tour on 28th March 1965. Although the goods yard was closed to traffic the previous year, the track work and most of the surviving goods facilities are still in place. Of particular note is the curious replacement goods shed, with its short unloading platform, and the long back siding, complete with its crane and coal staithes. The site was cleared and all the track-work lifted during the summer of 1965. The locomotive lasted slightly longer, being withdrawn from service in 1967.

The view in the down direction, towards Ockley, from the passenger foot bridge during the morning of 4th June 1969. To the left, the goods yard has been entirely cleared, apart from the short siding leading to the cattle dock, which is being used by the permanent way staff to berth their self-propelled trolley. However, in the distance, the trailing cross-over is still in place beyond the end of the up platform end, as is the up lay-over siding with its cat-walk.

This view in the up direction, towards Dorking and London, during the afternoon of 27th June 1969, was probably photographed from the down starting signal. The large station name board on the up platform is particularly prominent. Less obvious are the 'totem' name boards attached to the pre-cast concrete SR lamp posts. Of particular note is the removal of the cover to the footbridge. To the right, the engineer's department permanent way [PW] trolley is berthed in the erstwhile cattle dock road.

The Holmwood station building catches the morning sun on 23rd June 1976. Their proximity to the former A24 main road is worthy of note. During the winter, even at this late date, the Station Master, Mr Lee, kept a fire burning in the waiting room to greet travellers. Across the road, a cabin housed a small sweet shop, on the site of what is now Wren Cottage. This station building was demolished in April 1986.

In recent times earthworks on the line between Dorking and Horsham have started to fail, leading to wrong-line working or prolonged cancellations and some clever engineering solutions.

Top Left: On 5th May 1984 engineers have possession of the up line between Holmwood and Ockley whilst repair work takes place on an embankment.

Top Right: Also on 5th May 1984, an up train works wrong-line into Holmwood. Note the set of the points and the aspect of the dolly ground signal. The train will stop at the 'on' starting signal at the far end of the up platform

Left: On 8th May 2002, class 455 EMU in Connex South Central livery crosses a new embankment as it approaches Holmwood from Ockley. Other collapses took place in 2012 and 2013/2014.

A summer evening, probably in 1990, sees 4-EPB Unit No.5410 drifting to a halt at the Holmwood up platform. The '84' headcode indicates that this is a Horsham to Victoria train, via Mitcham Junction. Although the semaphore starting signals are still in place, the up lay-over siding and its associated cross-over have been lifted. As the signal box was 'switched out' during this period, the signals were permanently in the off position. Happily the signal arm nearest the camera was saved from the scrap-men and is now a part of the collection of railway artefacts in Dorking Museum.

An unusual visitor to Holmwood. Class 460, No.460001, an Alstom 'Juniper' type electric multiple unit designed for the Gatwick Express services, speeds through Holmwood on 22nd September 1999, whilst on test. Note the train's lack of an operating company livery and a 'nose-cone' fairing beneath the front windows of the cab. Also see how the hedge adjacent to the signal box has been allowed to grow over the years.

Inset: Class 460, No. 46002, complete with 'nose cone', passes through East Croydon, non-stop from London Victoria to Gatwick Airport.

115

Chapter 22
STEAM SPECIALS

Steam locomotives are still to be seen at Holmwood station. This chapter illustrates a very small selection of these special workings.
Above: On 30th May 1999, BR Standard Class 5 4-6-0, No. 73096, built at Derby in 1955, passes through Holmwood on one of the specials that shuttled back and forth between Horsham and Dorking that day to celebrate the arrival of the Reading, Guildford & Reigate Railway in Dorking in 1849.

On 20th June 2001 rebuilt West Country Class 4-6-2 No. 34016 'Bodmin' steams through Holmwood station, en route from Victoria [dep. 11-11] to Chichester [arr. 13.29]. This train was timed to make a 13 minute stop for water at Warnham at 12-15. The return journey from Chichester was via Basingstoke and Woking to London Waterloo.

On 5th July 2014, BR Britannia Class 4-6-2, No. 70013 'Oliver Cromwell', takes water at Holmwood whilst working the Horsham [dep. 08.30] to Canterbury West [arr. 13.09] 'Cathedrals Express', vîa Clapham Junction, Longhedge Junction, Clapham High Street and Tonbridge.

During this water stop at Holmwood, No. 70013 was blowing off vigorously and noisily from both safety valves whilst the footplate crew attempted to stop one of the injectors blowing through. This is the reason why steam is issuing forth at track level and the crew are peering anxiously over the cabside. Happily, order and peace were quickly restored.

On 2nd December 2015, Class B1 4-6-0, No.61306 'Mayflower', built by the North British Locomotive Company in 1948, is also seen taking water at Holmwood [note the orange coloured hose laid down the 1986 staircase leading to the up platform]. This working was the Horsham to Oxford 'Cathedrals Express', running via Clapham Junction, Kensington Olympia and Reading.

Above: A modern edition of 'The Type-Writer Girl' by Grant Allen, writing under the pseudonym of Olive Pratt Rayner, first published in 1897.

Left: 'Coldharbour from 'Roffy's'' by Lucien Pissarro, 1916

Chapter 23
EPILOGUE

Hitherto, this scrapbook has adhered to the 'history is one darned thing after another' approach and has been set out more or less as a continuous narrative. However, some events are not reliably fathomable, whilst others refuse to fit neatly into any paradigm at all.

In the former category falls the sojourn by Lucien Pissarro in Coldharbour, a small village in the hills above Holmwood. It is well documented that he stayed there with a Mrs Ansell between November 1915 and July 1916 and produced several notable works during that time, including 'Ivy Cottage, Coldharbour: Sun and Snow' [in the Tate Gallery] and 'Coldharbour from 'Roffy's', both signed and dated 1916. In July 1916, have becoming a naturalised British citizen, Pissarro left Coldharbour and moved to Sedgehill, near Shaftesbury in Dorset.

Yet, how did such a notable painter travel to and from Coldharbour? The village was not a recognised artists' colony, so where did essentials, such as canvases and paint supplies, come from and how were they delivered? Naturally, it would be joyous to record that Holmwood station was central to all this and that John Hoad was employed to carry the artist and his materials back and forth, up and down the hill. But unfortunately this is not to be, as researches have not yet unravelled this mystery – but the work continues.

An example of the latter category is 'The Type-Writer Girl', the 1897 novel by 'Olive Pratt Rayner'. In some quarters, this book is viewed as an icon of feminist literature and undoubtedly its main character is a 'new woman' of the late-Victorian period who, having attended Girton College, wears 'rational clothing', smokes cigarettes

A rare sight: passengers crowd the platforms of Holmwood station on a Sunday. During the 2012 London Olympic Games, Southern Trains used 10-coach class 377 'Electrostar' trains to run a free 30 minute shuttle service through Holmwood to allow spectators ready access to the cycling races that passed through Dorking and traversed Box Hill several times. This photograph was taken on 28th July 2012, en route to the Men's Road Race.

and travels about the countryside on her bicycle. However, the book was actually written by a Canadian man, Grant Allen. On the face of it, none of this is germane to the story of Holmwood station, but for the fact that a key scene in the novel is set at the station and it is also used as a symbol of a time of innocence at the dénouement of the plot.

As with 'The Battle of Dorking', it is quite apparent that the author of 'The Type-Writer Girl' was well-acquainted with the geography of the area between Dorking and Horsham and current events in England at the time. In the book, the niceties of etiquette create some difficulties for the two main female characters whilst attempting to buy tickets at Holmwood for themselves, their wrecked bicycles and a stoic lap dog, leading to his exchange of dialogue:

> "Then I can't go either", she cried, wetting her lips with fear. If you stop, I must stop with you, and telegraph for my father.
> I stared at her in astonishment "Why so?" I asked at last.
> "Why, because - because of this dreadful murder!"
> "What murder?" I inquired …
> She stared in turn. "You must have heard of it," she exclaimed. "It has been in all the papers." …
> She went on to explain to me that a woman had been found killed in a first-class carriage – stabbed to the heart, and stuffed under the seat – only three days before.

Almost certainly this is a reference to the murder of Miss Elizabeth Camp, whose body was found in a railway carriage at Waterloo station on 11th February 1897. Unsurprisingly, the facts in the novel and contemporary newspaper accounts are not entirely congruent, but an insightful comment made by a police spokesman at the time is repeated in a subsequent exchange between the same characters:

> "Have you reflected," I said drily, "that a woman may have committed that murder?"

Goodness me! Whatever next?

Yes. Whatever next, indeed? Well, let's think: trains on a Sunday at Holmwood would undoubtedly prove very useful both for visitors, with or without bicycles and well-behaved dogs, and local inhabitants alike; as would a return service in the evening that would allow visits to the cinema and other entertainments in Dorking or Horsham.

So, whilst it is on this hopeful thought that this scrapbook of snippets, photographs, newspaper cuttings and other ephemera comes to the end of its journey, it is not the finale of the Holmwood station story - far from it! Thankfully, trains will still continue to call at its platforms daily for the foreseeable future: except, of course, unaccountably on Sundays and during periods of equally inexplicable 'industrial action'.

Yet perhaps best of all, its rural Victorian origins continue to shine through with every arrival at the station, in the announcement:

The next station is Holmwood.

Passengers are requested to alight from the front five coaches of this train, as this station has a short platform.

APPENDIX 1

The 'Brighton Gazette', Thursday 16th October 1845 carried a lengthy notice at the top of page 2, setting out the intentions of the company, naming its supporters, the route of the line and how to apply for shares. This is a slightly abridged version of its text:

THE DORKING, BRIGHTON AND ARUNDEL ATMOSPHERIC RAILWAY, BY HORSHAM AND SHOREHAM. (WITHOUT A TUNNEL).

Capital, One Million, in 50,000 shares of £20 each

DEPOSIT £2.2S. PER SHARE

(Provisionally Registered, pursuant to the 7th and 8th Vict., cap 110.)

PROVISIONAL COMMITTEE

[There then follows a list of 80 names, including Stephen Dendy, Esq of Leigh Place, Surrey; Edward Kerrich, Esq of Arnolds, Capel, Surrey and Andrew Spottiswoode, Esq of Broom Hall, Dorking, Deputy Governor of the Union Bank of London, and Chairman of the Irish North Midland Railway.]

COMMITTEE OF MANAGEMENT

Chairman – Sir James Duke, M.P., and Alderman of London.

Deputy Chairman – Daniel Whittle Harvey, Esq.

Walter Wyndham Burrell, Esq. Captain Beare. William Collins, Esq, M.P. George Kirwan Carr Esq. William Fitzgibbon, Esq. Alderman Humphery, M.P. Richard Heaviside, Esq. Hugh Ingram, Esq. John Norton, Esq. Andrew Spottiswoode, Esq. Montagu David Scott, Esq. Thomas Watson, Esq.

James Whiting, Esq.

CONSULTING ENGINEER.

Charles Vignoles, Esq., F.R.A.S., M.R.I.A

ENGINEERS.

Messrs. Sherrard and Hall.

BANKERS.

The Union Bank of London, Moorgate Street

The London and County Joint Sock Bank, Lombard Street and Brighton: and at their several County Branches.

Messrs. Hall, West, and Boorer, Union Bank, Brighton.

SOLICITORS.

Messrs. Campbell and Witty, 21, Essex Street, Strand.

Messrs. Attree, Clarke, and McWhinnie, Brighton.

Messrs. Upperton, Verrall, and Veysey, Brighton.

Messrs. Coppard and Rawlison, Horsham.

Messrs. Everest and Wardroper, Epsom.

Richard Holmes, Esq., Arundel.

LOCAL AGENT.

John Dendy Sadler, Esq., Dorking.

SECRETARY, pro tem.

Robert Furner, Esq.

OFFICES.

The Adelaide Hotel, London Bridge.

This Company has been formed for the purpose of giving to the populous and wealthy district between Dorking, Shoreham, Brighton, and Arundel the advantage of a direct Railway communication with the Metropolis and the Coast, from which it is at present excluded.

It is proposed that the Line shall be constructed on the Atmospheric Principle, by which a cheap, speedy, safe, and frequent means of transit will be secured to the country through which it passes.

It may be observed that if any doubt has hitherto existed as to the efficiency of this Principle, that doubt has been most satisfactorily removed by the recent trials on the Croydon Railway, the results of which have exceeded the most sanguine anticipations, and prove that Railways constructed on the Atmospheric Principle must supersede all others.

A Railway on this Principle, from the Metropolis to Epsom, has already received the sanction of the Legislature: and, it is presumed, will be extended to Dorking, by means of the Direct London and Portsmouth Railway, a Bill for which has not only passed through all its stages in the House of Commons, but was most favourably reported on by a Committee of the House of Lords during the last Session; and although at the last moment it was unexpectedly re-committed, it will resumed in that stage at the commencement of the ensuing Session. In the event, however, of that Bill not passing, the promoters of this Line will be prepared to commence at Epsom, and to adopt either the Locomotive or Atmospheric mode of traction throughout the Line, as may be considered most advisable.

The Line (which has been examined by Mr. Vignoles) and found to be entirely free from engineering difficulties, will proceed from Dorking (or Epsom) to Horsham, and thence, in southern direction, by Shoreham to Brighton, where it is intended to have a Western Terminus; and from Horsham, in a South-Western direction, by Pulborough to Arundel, thereby affording an easy means of access to Littlehampton and Bognor.

The important district of country through which the Line passes is at present excluded from any direct means of Railway communication, either with the Metropolis or the Coast; and the level nature of the country renders the formation of the line a matter of comparative facility, without a large expenditure of capital; whilst the population and wealth of the towns and neighbourhood traversed, and the various means of intercommunication arising from Railways already in progress or contemplation, will produce such an amount of traffic as will insure an ample return for the capital required.

This Line will form the best connecting link between the district through which it passes, and the ports of Shoreham, Arundel, Portsmouth, and Southampton, and the principal towns of Surrey, Sussex, Kent, and Hants; and thus a readier facility will be afforded to the Traveller embarking for the Continent.

The consent of several of the chief Landowners on the Line has already been obtained, and the promoters have the satisfaction of stating that they do not anticipate any serious opposition from any persons whose property will be locally affected; but, on the contrary, they confidently rely on their co-operation in carrying out the project.

A detailed estimate of the expected revenue has not yet been made, but, from a calculation founded on the existing traffic, without the ordinary addition on account of Railway accommodation, no doubt can be entertained that a very large income will be realised.

Until an Act of Parliament shall be obtained, the affairs of this Company will be under the control of the Committee of Management for the time being, to whom power is given to allot the shares, and to apply the funds of the Company in payment of all the expenses incurred in its formation, and in the preparation of the plans and sections to be submitted to Parliament.

Power is reserved to the Committee to alter the amount of the proposed Capital, to vary the general course of the Line, and to make Branches therefrom, and to relinquish any portion of such Line or Branches and to enter into such arrangement with any other persons of Company as they may deem expedient.

In order to comply with the standing orders of Parliament, and to meet the preliminary expenses, a Deposit of £2 2s per Share must be made on Allotment.

Application for Shares to be made, in the subjoined form, to the Secretary, to any of the Solicitors, or to undermentioned Brokers, viz:- Messrs. George Burnand and Co., Cornhill; Messrs. Mieville and Co., Angel Court, Throgmorton Street: Messrs. Marten and Hezeltine Finch Lane Cornhill; Messrs. Whitmore and Sons, Change Alley, Cornhill; Messrs. Pownall and Worthington, Liverpool; Mr. James Bowden, Hull; Messrs. Cardwell and Sons, Manchester.

October 4 1845.

FORM OF APPLICATION FOR SHARES.

To

The Committee of the Dorking, Brighton and Arundel Atmospheric Railway, by Horsham and Shoreham.

I request you to allot me   shares of £20 each in the above undertaking, and I hereby undertake to accept the same or any less number which may be allotted to me, and to pay the deposit thereon, and to execute the Parliamentary Contract and Subscribers' Agreement when required.

Dated the   day of   1845.

Name in full ....

Residence ....

Business or Profession (if any) ....

Reference ....

APPENDIX 2

On Saturday 30th November 1861, amongst several other similar notices placed by the promoters of other proposed railways, the 'Sussex Express' carried a prolix notice buried in the middle of page 8, setting out the intention of the Horsham, Dorking and Leatherhead Railway Company to apply to Parliament for an Act that would allow the railway, "with all proper works and conveniences connected therewith" to be built. This is the entire text of that notice, describing the route and many other legal or technical details:

HORSHAM, DORKING, AND LEATHERHEAD RAILWAY

(Incorporation of Company for making Railways from Horsham to Dorking, and from Dorking to Leatherhead: Traffic arrangements with London and South Western, London, Brighton, and South Coast, and South Eastern Railway Companies; Power to those Companies to subscribe and Amendment of Acts).

NOTICE IS HEREBY GIVEN, that Application is intended to be made to Parliament in the ensuing session for an Act to incorporate a Company for making and maintaining the Railways following, or one of them, with all proper works and conveniences connected therewith and approaches thereto respectively [that is to say]:

FIRST: - A Railway commencing in the Parish of Horsham, in the County of Sussex, by a junction with the London, Brighton, and South Coast Railway, at or near the Northern end of the Horsham Station thereon, and terminating in the Parish of Dorking, in the County of Surrey, by a Junction with the Reading, Guildford, and Reigate Line of the South Eastern Railway Company, at or near the Eastern end of the Boxhill Station thereon, and which intended Railway and works will pass from, in, through, or into, or be situate within the parishes and extra parochial or other places following, or some of them (that is to say): Horsham, Rusper, and Warnham in the County of Sussex, and Capel, Ockley, Newdigate, Charlwood, Leigh, Betchworth, and Dorking, in the County of Surrey.

SECOND: - A Railway to be situate wholly in the County of Surrey, and commencing in the said Parish of Dorking by a Junction with the first described intended Railway at or near the point where the same is intended to Cross the Turnpike Road leading from Dorking to Reigate, 300 yards or thereabouts to the eastward of the "Punch Bowl" Inn, in the said Parish of Dorking, and terminating in the Parish of Leatherhead by a Junction with the Epsom and Leatherhead Railway at or near the Terminus thereof, which said intended Railway and Works will pass from, in, through, or into, or be situate within the several parishes and extra-parochial or other places following or some of them (that is to say): - Dorking, Mickleham, Fetcham and Leatherhead, all in the County of Surrey.

And it is proposed by the intended Act to empower the Company to be thereby incorporated to purchase by compulsion and by agreement Lands, Houses and Hereditaments for the purposes of the proposed Railway and Works, and to alter, vary or extinguish all existing rights and privileges connected with such Lands, Houses and Hereditaments, or which would in any manner interfere with the construction, maintenance, and use of the said proposed Railways and Works, or any of them, and to confer other rights and privileges.

And it is intended by such Act, to take power to stop up, cross, divert, or alter, either temporarily or permanently, any turnpike or other roads, streets, highways, bridges, footpaths, ways and rights of way, railways, tramways, canals, aqueducts, rivers, navigations, streams, pipes, sewers, drains and water courses within the said parishes and extra parochial or other places, or any of them, which it may be necessary to stop up, cross, divert, or alter, for the purpose of the said intended Railways and Works, or any other purposes of the said Act.

And it proposed by the said intended Act to take powers for levying tolls, rates, and duties for or in respect of the use of the said proposed Railway and works, and the conveniences and accommodations connected therewith and to confer, vary, or extinguish exemptions from the payment of such tolls, rates, and duties respectively.

And it is also proposed by the said intended Act to enable the Company to be thereby incorporated, and the London, Brighton, and South Coast, South Eastern and the London and South Western Railway Companies, or any or either of such Companies, to subscribe or contribute funds towards the construction and maintenance of the said intended Railways and Works, or either of them or any part or parts thereof, and to guarantee such interest, dividends, annual or other payments in respect of the monies expended in the Construction thereof, as may be agreed upon between such Companies respectively, and to take and hold shares in the Capital of the Company, and to apply to the purposes aforesaid, or any of them, any Capital or Funds, now or hereafter belonging to them respectively, or under the control of their respective Directors, and if they shall think fit to raise additional monies for that purpose, by the Creation of New Shares in their respective Undertakings, with or without preference or priority or other rights or privileges, or by mortgage, or bond, or by both those means, or by such other means as Parliament shall authorise and direct.

And it is proposed by the said intended Act so far as may be necessary for the purposes aforesaid, to alter, amend, enlarge or repeal the powers and provisions of the acts following, or some or one of them, relating to the London, Brighton and South Coast Railway Company namely [then follows a list of statutes] and any other act or acts relating directly or indirectly to, or affecting the, London, Brighton, and South Coast Railway Company.[Similar provisions are then made for the London & South Western and the South-Eastern Railway Companies].

And notice is hereby further given that on or before the 30th day of November instant, Plans and Sections of the proposed Railways and works, and a Book of Reference to such Plans, together with a published Map with the Line of the proposed Railways delineated thereon, a copy of this Notice, as published in the London Gazette, will be deposited for public inspection with the Clerk of the Peace for the County of Sussex, at his Office at Lewes, in the said County, and with the Clerk of the Peace for the County of Surrey, at his Office, at Lambeth, in the said County, and that on or before the said 30th day of November, a copy of so much of the said Plans, Sections, and Book of Reference as relates to the each Parish or extra parochial place in or through which the intended Railway and Works will be made, or in which any lands intended to be compulsorily taken are situate, and a Copy of this Notice, as published in the London Gazette, will be deposited for public inspection in the case of each such Parish, with the Parish Clerk thereof, at his residence, and in the case of each such extra parochial place with the Parish Clerk of some Parish immediately adjoining thereto, at his residence, and that on or before the 23rd day of December next, printed Copies of the Proposed Bill will be deposited in the Private Bill Office of the House of Commons.

Dated this 18th day of November 1861,

W. Gascoigne Roy, 25, Great George Street, Westminster, Solicitor for the Bill

APPENDIX 3

The Daily News, Thursday 1st August 1872 - Boarded Out

The great Union school for the pauper children is a fine sight. Its Gothic front adds a new beauty to the landscape. Its dining-room is a master-piece of chamber architecture. Its wards, for the stately discomfort about them, might be the sleeping apartments in a palace, and the life within is a fine life. The children feast by ring of bell, like great lords and ladies, and get their mental and moral training by the stroke of the clock. At 10 a.m. they are learning to be good; at half past 11 to be wise. They promenade like an army on the march; and advance on nature in the line formation. So much for the best of the pauper schools - for those which are separated from the dreadful poor-houses to which they belong by many miles of sweet heather and of breezy downs. Of the worst, which are properly speaking not schools at all, but mere infants' wards in those dismal unions of the great towns in which the poor of all ages are interned, we cannot trust ourselves to speak. The children contaminate each other both in mind and body, and are prepared for permanent pauperism, or worse. Happily, however, these establishments will soon belong exclusively to the history of the past. It is by their new Union School that our local authorities must stand or fall. But there is something wanting in the Union School - perhaps a little less of the logic of social life. If the labourer's calling, to which they are most of them destined, really were in the nature of a battle and march the Union School would be the very thing for the pauper boys and girls. But few of them in after days will be lucky enough to have their wants supplied by a commissariat and their initiative by a word of command. Their rule of life will have to be a more or less scientific adaptation of the rule of thumb; and that has no place in the course of study in these magnificent institutions. Of the kind of education that is to be found in the rule of three they will get plenty, but of that which is received through the medium of family joys and sorrows none at all. Home partings and meetings, festivals and mourning; the housewifely sense of treasure in humble stores; the joint pride of man and woman in the bravely earned and as wisely spent will be as things unknown to them. They will come out of the school with a quaint conception of the world as a place in which even the blessings are served out in rations share and share alike to all. And there is another consideration which perhaps weighs more with the guardian mind than any of the above - the Union School is dear, for when all accounts are settled the average cost is nothing less than £17 per annum for every child.

For the last year or two the English guardians have been cautiously trying the system of boarding-out, which, as it has been found to answer in Germany and in Russia – not to speak of Scotland and of Ireland – comes to us with the needful weight of foreign recommendation. The first object of the system is to supply the influences of home. Decent cottagers are found in the country and the children are sent to live with them. They go to school and to church, they assist in the housework under proper conditions, and in every respect are treated as members of a well-conducted labourer's family. The clergyman and his wife, and any number of other ladies and gentlemen, carefully selected, look after them, and make periodical reports on their condition; and, for what must after all be the supreme test of the system, the cost - it averages just

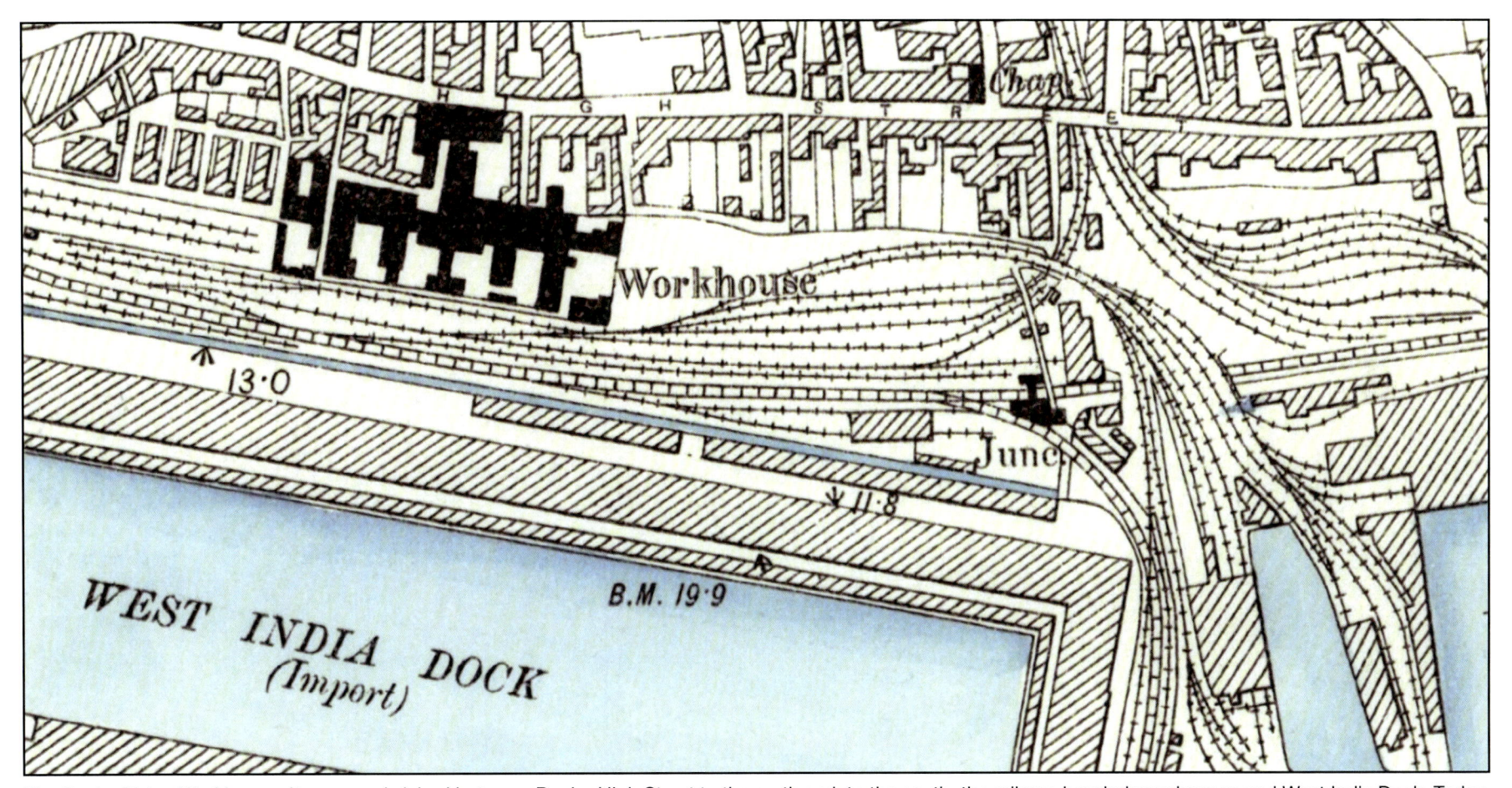

The Poplar Union Workhouse. It was sandwiched between Poplar High Street to the north and, to the south, the railway, bonded warehouses and West India Dock. Today, the site forms part of the Canary Wharf office and shopping complex.

£10.10s. against the £17 quoted above. From time to time parties interested in the welfare of the boarders have published reports on the working of the new scheme. In January last the Hon. W. Warren Vernon and Col. Fremantle went to Culverton and St. Mary's Wolverton, in Buckinghamshire, and gave the most cheering account of some children sent there by the guardians of St. George's. Equally weighty testimony came from the same quarter as to other parts of the country. There are children at Fox Warren, Cobham, and Ripley, in Surrey, and they are all doing well. There are also children at Holmwood, in the same county, and these the writer went to see the other day, and to report for himself.

Holmwood is close to Dorking, and Dorking, as everybody knows, is very nearly the loveliest place in the south of England. The traces of the late dreadful battle have entirely disappeared. The

'A Dame's School' by Thomas Webster 1845 © Tate, London 2017

vicarage is intact, and still commands a view of twenty miles of corn and woodland; the high road, so dreadfully cut up by the passage of the invader's artillery, is once more as smooth and as firm as a park walk; and the bye-lanes branching off from thence into the fields show the same unbroken lines of hedgerows as they did before the invader came. In one of these bye-lanes not far from the Holmwood station, the writer, being under clerical guidance, got on the track of two infant boarders out. The cottage, which was their new home -exchanged for a ward in the Poplar Union - might have been higher and broader with advantage, but it could not have been cleaner, or - what is something to the purpose - more picturesque.

It was delightfully old-fashioned, and had an unmistakeable chimney corner, with a seat ready for the story teller and a shelf for the listener's pipe and mug. No one was at home but the housewife, a comely person, not without that look of resignation to mild suffering which is suggestive of a family of children "on the mind". As she was entirely free from the affectation of affection for her pauper charges, it is probable that she possessed the reality of it. The little ones were at school, she told us, with the "other children" (her own). Would we go and see them? They were quite well. We did not ask about their happiness, for with the school-house at hand we were in a position to obtain more conclusive evidence than hers on that point. The visit had, however, made this clear: The cottage was a place in which a child might be very happy indeed. The school was in all essentials that of Webster's famous picture.

The building, having been erected for its present purpose, had certainly a "Government" look about it, but the children were as delightfully irregular as ever. One read his book, another thumbed it merely, while a third toyed with the forbidden apple in the corner out of the reach of the teacher's eye. The first room we entered was for the infant class of both sexes, and among the infants was one of the boarded-out girls from the cottage. As they were all, without exception, ruddy and cheerful, it seemed clear that any child of their number must be healthy and happy too. But presently, on some kindly pretence of asking a question about her books, the little girl was beckoned forth; and there stood a kind of acknowledgement in full of all reasonable demands in regard to cleanliness, plumpness, and perfect childlike content with self and the world. The blush that mantled upon her cheek as she faced the writer called upon another to his own, as he felt that he was expected to be a Bishop about to offer a prize, and knew that he was nothing of the sort. In the next school-room was a mixed class of older children; and here were two more of the boarded out – the third hailing from a cottage we were shortly to visit. A schoolmaster was in charge, and he was giving them a lesson on "Bears". The dimensions, habits, tastes, and tricks of Bruin were all set down on a large slate, and that information imparted, the slate was turned around to the wall, and the children catechised from memory. "What does he live on?" asked the master. A dozen

A performing bear, watched by a youthful audience in Dorking High Street, c.1895.

Holmwood School and Pupils

Left: c.1890 Right: c.1905. Between the dates these two photographs were taken, the turnpike has been improved and a pavement added.

arms were advanced to show that as many pupils were ready with the information. One was selected; and forth, with all the promptitude of certainty, came the answer, "Honey". "Anything else?" asked the master, and again the arms went to work as a preliminary to the statement, "Bees". "Where do they live?" was the next question. "The white ones live in the cold", said a young gentleman who appeared to have made natural history is too exclusive study. "Are bears ever met with anywhere else?" asked the vicar, and there was a sly twinkle in his eye. "Yes, Holmwood", answered a little girl, with a sympathetic glance of fun. It was the vicar's joke - an unfortunate bear who had been kidnapped into the theatrical profession having been marched through the village sometime before.

Two of the boarded out were, as we have seen, in this room, but as no questions happened to be addressed to them, they had no opportunity of showing their proficiency in the studies of the day. Their hands were often enough held out in a token of readiness, but, like many other persons of possible distinction in the world, they did not get their chance. And it is grievous to be obliged to say that the only distinctive circumstance in their conduct was a furtive attempt at "tickling in class" during the investigations concerning the white bear. The master did not see it, and it was not reported to him, for it was felt that however little it had to do with natural history, it supplied some information germane to that branch of the inquiry into human happiness which was the writer's study at Holmwood on this particular occasion.

A little way beyond the village school lay the cottage in which the third boarder lived. It was like the first, cleanly and comfortable, and it contained that most picturesque piece of cottage furniture, an aged peasant enjoying the repose of his own fireside. The housewife here was a woman of the severely tidy type. Craftiest spiders, it was evident, had no chances against her, and tables and chairs seemed to be standing at attention in her presence – nay, the very peasant looked as if he had been dusted all over before he was placed in the ingle nook. She was hopeful about her boarder, but she would not permit herself to be enthusiastic. The child was too old, eight or nine, when she came from Poplar. She knew nothing of the rudiments of house work. Many of her habits were formed, and she was sometimes not easy to manage. She was not dirty when she came, and yet she wasn't clean. "It was the corners like", if the housewife might so express herself. At this the peasant gave a short, dry cough, indicative on the surface of his manner of assent, but, beneath the surface of very much more. A walk of a couple of miles or so along the high-road led to another school, nearer to Dorking, in which might be heard tidings of the last four girls boarded out in the neighbourhood. It was the dinner hour, and the children had been dismissed, but the school-mistress very obligingly showed the way to the college [sic] (cottage?) in which this pupil lived. The pupil was at play in the garden. The housewife went out to fetch her, leaving a small representative of her own race to watch the stranger suspiciously through the bars of a chair. After a moment's interval for "tidying" the boarded-out appeared. Preparation might have had something to do with the clean pinafore, and coquettishly fastened hat, but it could not have given the apple-like redness to the cheeks, or plumpness to the limbs. The boarded-out was abashed and spoke not, and showed a disposition to cling to the foster-mother's dress. If the inquirer had before felt sorry that he was not a Bishop he now regretted that he should be mistaken for a gipsy. To his questions as to age and name, the fondness for pudding and play, meant to ne kindly, the boarded-out vouched no word: and when he repeated them, with an insistence that was meant to be jocular, he at length received for reply only an inarticulate gurgle that might have signified yes or no. The evidence here, in short, was purely ocular, not oral; but it was very conclusive of its kind, for it showed a long course of that motherly care which must always have for its motive motherly affection.

So ends the amateur report. Taken with those which have already emanated from official sources, it may seem conclusive as to the success of the partial experiment of boarding out. The children who have had the advantage of this twofold training of home and school must surely be better qualified for success in life than those who have seen the world only through the medium of an "institution". Of the guarantees for the good behaviour of the foster parents, it is not needful here to say more than that they seem absolutely perfect. To enable a child to be ill-treated, not the cottagers alone, but the clergyman of the parish and his wife, the schoolmaster and the squire, with any number of the ladies of local magnates who form the Boarding-out Committee must be in a vast conspiracy against its happiness, with the inspectors of the Board of Guardians favouring their designs. It only remains to add, that those who desire to know more of the system as apart from its obvious results may the requisite information of the Secretary of the Howard Association in Bishopsgate-street Without.

APPENDIX 4

An abstract of statistics derived from the paper, "The Reparation of Betchworth Tunnel, Dorking, on the London, Brighton and South Coast Railway", by George Lopes BA (Camb.), Assoc. M. Inst. C.E., published in the Minutes of the Proceedings of the Institution of Civil Engineers, Volume 95, in 1889:

- Work on the 55 yards of new inverted section in the slip took place between August 8th and December 30th 1887, and on the remaining 330 yards between November 7th 1887 and February 7th 1888. It was continuous, day and night, with the exception of Christmas-day.

- In the reconstruction, one of the two lines of railway through the tunnel was taken up and the other slewed to the centre for the convenience of the contractor, who used a six-wheeled locomotive which worked through the tunnel.

- In addition, during the busiest part of the reconstruction, six horses and three hundred men were employed.

- Two and one-third million bricks were used. They were the best Horsham stocks.

- For mortar, 780 tons of cement was supplied by the Sussex Portland Cement Company, Newhaven. It was mixed with 2,000 cubic yards of sand, some of which came from pits at New Cross and the remainder from beds at Oxted.

- At first, water was provided in tanks from Horsham and Holmwood. Subsequently, a 2-inch pipe from the mains supply from the Dorking Water Company was laid through the tunnel, with junctions and bib-cocks where necessary.

- The following materials were supplied: 62 elm skeleton ribs; 114 segmental centres; 16,500 cubic feet of timber; 3,030 cubic feet of larch; 7,400 lineal feet of 3-inch by 7-inch battens; 40,600 lineal feet of 3-inch by 9-inch deals ; 52¼ fathoms of poling-boards; 48,700 page-, driving-, and raking-wedges; 760 slack blocks; 10 tons of hay; 10,525 gallons of naphtha and 3,474 lbs of candles.

- The slip comprised about 33,000 cubic yards.

- The cost of the work per lineal yard of the inverted section was £145, and of the relieving-arch section £30. The exceptionally heavy expense of the former was caused by the enormous quantity of timber necessary to support the slip, and to the necessity of building in so much of it.

- The tunnel was opened for traffic on the 1st of March, 1888.

APPENDIX 5

An extract from the 1857 plan showing the proposed route of the Shoreham, Horsham and Dorking Railway as it passed through the village of Newdigate. The railway line would have run to the east of the parish church of St Peter and between Dean House Farm and Horsielands Farm.

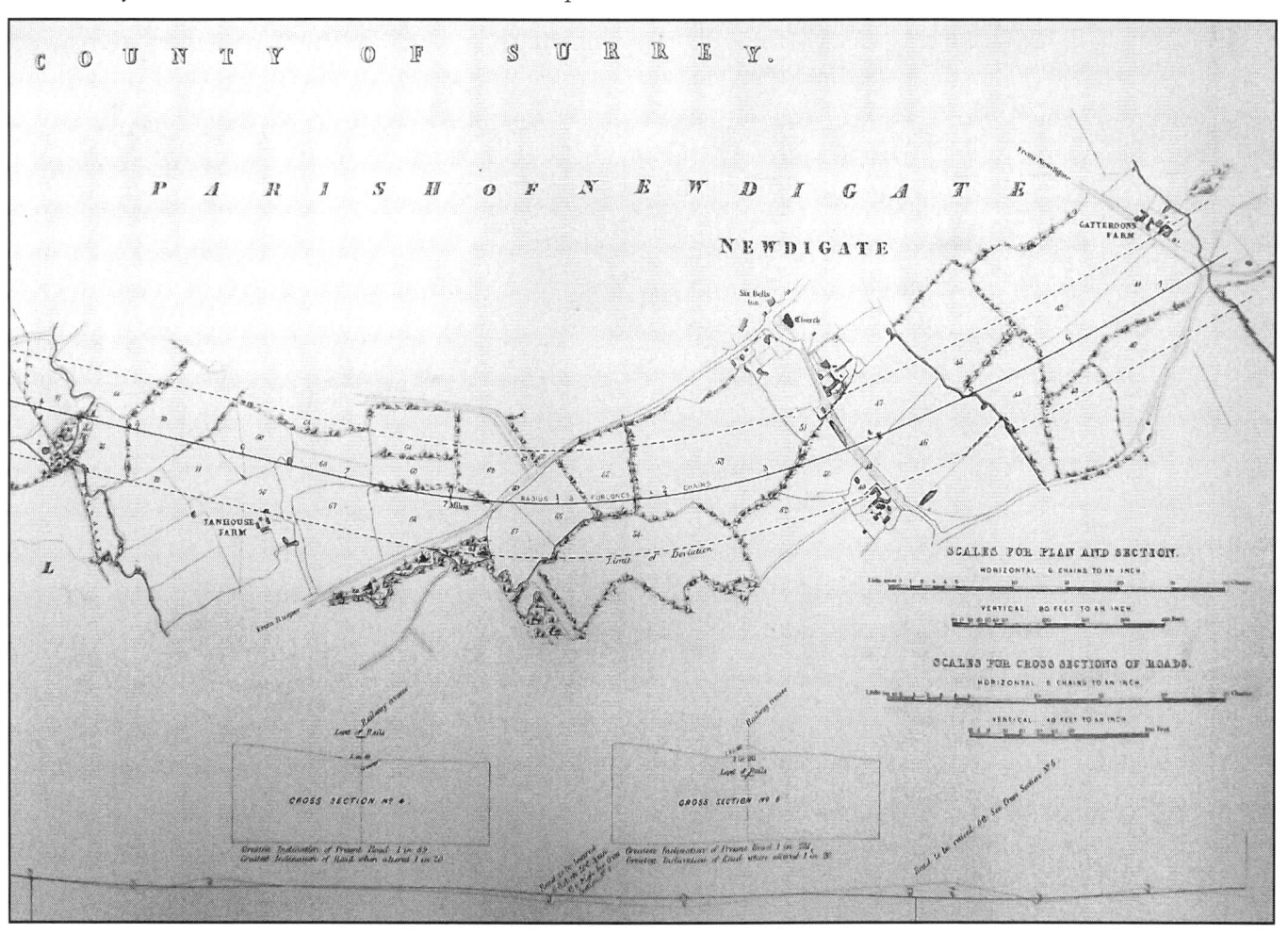

APPENDIX 6

The allocated numbers, formation and working restrictions for the different rakes of vehicles provided by a variety of railway companies for the transport of wounded service personnel, as listed on the back page of LB&SCR Ambulance Train Notice No.1, published on 5th March 1918.

9 FORMATION AND RESTRICTED WORKING OF AMBULANCE TRAINS

Train No.	Berthed at	Formed of	Brake
1	Southampton	10 G.C. Co.'s Vehicles	Dual
2	Southampton	10 G.C. Co.'s Vehicles	Dual
3	Dover	10 G.E. Co.'s Vehicles	Dual
4	Dover	11 G.W. Co.'s Vehicles	Dual
5	Dover	11 G.W. Co.'s Vehicles	Dual
6	Southampton	10 L. & Y. Co.'s Vehicles	Dual
7	Dover	10 L. & N.W. Co.'s Vehicles	Dual
8	Dover	10 L. & N.W. Co.'s Vehicles	Dual
9	Southampton	10 L. & N.W. Co.'s Vehicles	Dual
10	Southampton	10 L. & S.W. Co.'s Vehicles	Dual
11	Southampton	10 Midland Co.'s Vehicles	Vacuum
12	Southampton	10 Midland Co.'s Vehicles	Vacuum
13	} Ireland	—	—
14			
15	Southampton	10 G.C. Co.'s Vehicles	Dual
16	Southampton	11 G.W. Co.'s Vehicles	Dual
17	Southampton	10 L. & Y. Co.'s Vehicles	Dual
18	Dover	10 L. & N.W. Co.'s Vehicles	Dual
19	Dover	10 L. & N.W. Co.'s Vehicles	Dual
20	Dover	11 G.W. Co.'s Vehicles	Dual
21	Southampton	10 L. & S.W. Co.'s Vehicles	Dual
22	Southampton	10 G.E. Co.'s Vehicles	Dual
23 A.	Dover	9 S.E. & C. Co.'s Vehicles	Vacuum
26 A.	Southampton	7 L. & S.W. Co.'s Vehicles, including Dining Saloon	Vacuum
31 A.	Dover	11 L. & N.W. Co.'s Vehicles	Vacuum
37 A.	Southampton	6 L. & N.W. Co.'s Vehicles and 4 L. & S.W. Co.'s Vehicles, including Dining Saloon	Vacuum
38 A.	Southampton	8 L. & N.W. Co.'s Vehicles and 2 L. & S.W. Co.'s Vehicles, including Dining Saloon	Vacuum
39 A.		10 L. & S.W. Co.'s Vehicles, including Dining Saloon	Vacuum

Trains marked "A" are Emergency Trains.
Trains Nos. 26, 37, 38 and 39, if run via Crystal Palace, must be restricted to Single Line working through Crystal Palace Tunnel.
Trains Nos. 6, 37, 38 and 39 will *not* be used for services to the South Eastern & Chatham Railway.

ACKNOWLEDGEMENTS

To all the many people and organisations that have contributed time, effort, material and memories during the compilation of the manuscript for this book, I am most truly grateful. However, my special appreciation goes to the following:

Bobbie Roundthwaite and Lenka Cathersides, who provided able assistance in the archives at Dorking Museum, and Clare Flanagan who reassured me about the well-being of the Holmwood up starting signal arm in the museum store. Support has come from both the Dorking Local History Group, in particular Gwen Wood and Martin & Maureen Cole, and the Newdigate Local History Society, especially John Callcut. In respect of the New House Farm material, I am much obliged to Carrie Crutcher for allowing its use and to Jane Lilley for reducing this huge resource to manageable proportions. Technical queries have been answered by the National Railway Museum, York, with informative responses from Joe Lane at their Search Engine facility, and by the Bluebell Railway, with Tony Hillman, Roger Cruse and Fred Bailey being particularly helpful in finding photographs and answering a perplexing problem. Capel local historian, Mary Day, has supported this project from the outset and was particularly accommodating regarding the compulsory acquisition of the Holmwood station site. Mary Hustings and Lady Wedgewood kindly added to my knowledge of Lucien Pissarro in Coldharbour. In the House of Lords library, whilst researching the various private Bills, all the staff encountered there were patience and consideration personified. Vital IT care was supplied by Chris Ball and his merry band at Helpdesq – without them, this book might have been lost forever.

Other useful and timely assistance has been kindly provided by Geoff Allen; Jennie Barnes; Nick Beck; Brian Buss; Peter Collis; Clive Coward and David Thompson at the Tate Gallery; Andrew Currie at Bonhams; Ian Dockra; Bob Fowke at YouCaxton; J J Heath-Caldwell; Eric Hoad; Zan Horne; Barry Hutt; Clare Laker; Jenny Lester; Andrew & Debs Mansfield; Mark McFadden; Vic Mitchell; Jan Morgan; Dr Bruce Osborne; Marion Roberts; John Russell; Richard Salmon; John Scrace; Graham Stacey and Mervyn Young.

As a wonderful source of artefacts and ephemera I cannot praise eBay, or the sellers that use it, too highly: this book would be much the poorer without the services that they have provided. A further debt is owed to all those people, both living and dead, who bothered to keep this material intact, rather than just simply throw it away. Without those scraps of paper and card, there would be no authentic evidence worthy of research and recovery.

Finally, three people need to be singled out for their particular help and generosity. The first is Lorraine Spindler, who has patiently taught me how to gain access to, and then use, the archive records that are available in the public domain. She also opened up my eyes to the possibilities of using such records to determine the provenance of photographs and to identifying the subjects appearing in them. The second is local historian and Holmwood resident, Kathy Atherton. Her general contribution to this project has been immeasurable, not least because she, to use her own words, "recognises the difference between a locomotive and a train". Last, but by no means least, is my wife, Lindsay. She has had to put up with tales of yesteryear [and, it seems, not all of them were interesting] for far too long, whilst our home slowly crumbles around us.

The errors, misunderstandings and omissions are all my own, as are the various opinions expressed in the text. J.S.W

PHOTOGRAPHIC & OTHER CREDITS

Acknowledgement is made to the following photographers or keepers of collections for the use of their illustrations in this book:

Dr I C Allen: page 81;
R K Blencowe: Frontispiece
Bluebell Railway Museum Archive/Alan Postletwaite: page 105;
Bonhams Auctioneers and Valuers/Andrew Currie: page 121 [left];
John Brown: pages 113 [all], 115 [main image], 116 & 117;
H C Casserley: page 87 [bottom];
Peter Collis; pages 118, 119 & 120;
Mary Day: pages 19 [all] and 20;
Ian Dockra: page 114;
Dorking Museum: pages 35 [right], 42 [left], 67, 72, 79 [top left], 90 [all] & 131;
Lens of Sutton: pages 64 & 80;
Newdigate Local History Society: page 100 [top right];
John Scrace: pages 108, 109, 110, 111 & 112;
John Scrace Collection: pages 22 and 23;
Bredan Sewell: page 47 [left];
Signalling Record Society/Scrimgeour Collection: page 107;
Southernposters.co.uk/ Joe Ellis: page 83 [left];
Surrey History Centre: page 15 [left];
Tate Images: page 130;
H. Gordon Tidey: page 37;
Wikipedia/Wikimedia Commons: pages 26, 49, 58 [all], 61, 70 [bottom right] & 82;
Mervyn Young Collection: page 42 [left].

The extracts from the Ordnance Survey plans reproduced on page 21 [published 1873]; page 92 [published 1914]; page 99 [published 1938] & page 129 [published 1896] are reproduced by the kind permission of the Customer Service Centre, Ordnance Survey Ltd. These extracts have not been reproduced true to scale.

Otherwise, all other images, photographs, pictures and ephemera are taken from the author's private collection. This has been acquired from a variety of open market sources over many years, often with no indication of its provenance. Whilst particular efforts have been made to determine the ownership of copyright and to give due credit, it is possible that material has been inadvertently published without consent. In the unlikely event that such an instance arises, the publishers will be pleased to hear from and individuals or bodies who may be affected in order that suitable corrections may be made in future editions of this work.

BIBLIOGRAGPHY

Anonymous - Locomotives of the London Brighton & South Coast Railway, The Locomotive Publishing Company Ltd London 1903
Atherton, Kathy - The Lost Villages: A History of the Holmwoods, K Atherton Dorking 2008
Brown, Antony - Cuthbert Heath: Maker of the Modern Lloyd's of London, David & Charles Newton Abbot & London 1980
Chesney, Sir George Tomkyns - The Battle of Dorking [Fiction], William Blackwood & Sons Edinburgh & London 1871
Course, Edwin - The Railways of Southern England: The Main Lines, Batsford London 1973
Day, Mary & Ettlinger, Vivîan - Capel: The Chapel by the Spring, Ammonite Books Godalming 2015
Green, Frederick Ernest - The Surrey Hills, Chatto & Windus London 1915
Hadfield, Charles - Atmospheric Railways, Alan Sutton Gloucester 1985 [reprint of 1967 original]
Hamilton, Genesta - A Stone's Throw: Travels from Africa in Six Decades, Hutchinson Ltd, London 1986
Haresnape, Brian - Railway Liveries 1923-1947, Ian Allen Ltd, Shepperton, Surrey 1989
Harrod, John T A - Up the Dorking, Southern Railways Group 1999
Horne, Alistair - The Fall of Paris: the Siege and the Commune 1870-71, Messrs McMillan & Co Ltd London 1965
Jackson, Alan A - Dorking's Railways, Dorking Local History Group 1988
Keat, Peter J. - Goodbye to Victoria The Last Queen Empress: The Story of Queen Victoria's Funeral Train, The Oakwood Press, Usk 2001
Mais, S P B - Southern Rambles for Londoners, The Southern Railway 1931
Malden, H E [Editor] - A History of the County of Surrey: Volume 3, Victoria County History, London, 1911
May, Walter M & Coaten, Arthur W - Thomas' Hunting Diary 1905-1906, Messrs Thomas & Sons, 32 Brook Street, W. 1905
Minnis, John - Railway Signal Boxes: A Review, Research Report Series No. 28-2012, English Heritage 2012
Mitchell, Vic & Smith, Keith - Southern Main Lines: Epsom to Horsham, Middleton Press, Midhurst 1986
Morris, O J - The Birth of the Southern. Trains Annual 1953, Ian Allen, London 1953
Newberry, Pat - 'Est-ce que vous savez le col d'arbor?', Abinger & Coldharbour Parish News March & May 2000
Norris, W S - Steam Days in Southern England. Trains Annual 1954, Ian Allen Ltd, London 1954
Rayner, Olive Pratt [pseudonym of Allen, Grant] - The Type-Writer Girl [Fiction], C A Pearson London 1897
Willox, W A & Lee, Charles E - Queen Victoria's Funeral Journey, The Railway Magazine March 1940. The Railway Publishing Company, London 1940
Womersley, Julian - The Surrey Union Hunt: Our History Unbuttoned, Edgebury Press Crawley West Sussex 2007

OTHER SOURCES

Internet:
General
https://historicengland.org.uk/listing/the-list/list-entry/1376781 [Holmwood signal box]
http://www.jjhc.info/helshamjoneshenry1920.htm
http://www.lbscr.org/
http://www.semgonline.com/steam/lbscr/lbscrlocos.html
https://en.wikipedia.org/wiki/London_and_Brighton_Railway
https://en.wikipedia.org/wiki/Leopold_Heath

Newspapers
The contemporary sources, as named in the text, are all derived from 'British Newspapers 1710-1953' at http://www.findmypast.co.uk/

UK Railway Employment Records
Years 1833-1956 at http://www.ancestry.co.uk:
LB&SCR Traffic Department Staff Book 1865-69
LB&SCR General Manager's Registers of Staff for 1870-74; from 1875; from 1885 and 1895
LB&SCR List of Station Staff – Holmwood for 1871, 1877, 1881 and 1891
LB&SCR Register of Appointments 7th May 1867

Parliamentary Archives:
House of Lords
HL/PO/PB/3/plan1845/L46 - London and Portsmouth Railway. Map and Section of the L&P Railway, through Dorking, Horsham and Arundel, etc
HL/PO/PB/3/plan1846/D17 - Dorking Brighton and Arundel Atmospheric Railway. Plan and Section, etc
HL/PO/PB/3/plan1857-1858/S8 - Shoreham, Horsham and Dorking Railway Plan, Section, Book of Reference, Published Map, Gazette Notice
HL/PO/PB/3/plan1862/H5 - Horsham, Dorking and Leatherhead Railway. Plan, Section, Book of Reference, Published Map, Gazette Notice, Lists of owners, lessees and occupiers, etc

The National Archives:
RAIL 414/570 LBSCR Books & Records, Special Traffic Notices

Private Papers:
New House Farm, Newdigate: The farm bailiff's daily diary [1908-1924]; the farm ledger [1903-1918] and the wages book [1917-1926].

ABOUT THE AUTHOR

Because of his naturally grumpy and misanthropic disposition, few have ever dared to ask author, Julian Womersley, quite why he has doggedly commuted from Holmwood station for so long. But whatever the reasons for this persistence, and they have changed over the years, they enable him to pursue a passion for railways, local history and country life in general.

He is a prolific writer, this is not his first book by any means, and other work has appeared in publications as diverse as *Horse & Hound*, *Farmer's Weekly*, *The Birmingham Post* and *The Estates Gazette*, or can still be heard on BBC Radio 4. His unexpectedly epic series of railway locomotive "road tests" was a feature in the now defunct *Steam Classic* magazine for several years. Even more improbably, one of his films once became a cult classic at the Royal Agricultural College - until the videotape wore out.

Although now partially retired, he remains a Fellow of The Royal Institution of Chartered Surveyors and continues to specialise in the resolution of disputes caused by arcane legislation.

Had the Dorking, Brighton and Arundel Atmospheric Railway ever been built, its trains would have been plainly visible from Julian's study. Happily, the sounds and smells of the steam locomotives that still regularly visit Holmwood waft in through its windows instead.

Author photograph by Colin Jacks